From Waste to Energy

From Waste to Energy

Technology, the Environment and the Implications under EU Law

Harry Post (Ed.)

eleven
international publishing

Published, sold and distributed by Eleven International Publishing
P.O. Box 85576
2508 CG The Hague
The Netherlands
Tel.: +31 70 33 070 33
Fax: +31 70 33 070 30
e-mail: sales@elevenpub.nl
www.elevenpub.com

Sold and distributed in USA and Canada
International Specialized Book Services
920 NE 58th Avenue, Suite 300
Portland, OR 97213-3786, USA
Tel.: 1-800-944-6190 (toll-free)
Fax: +1 503 280-8832
orders@isbs.com
www.isbs.com

Eleven International Publishing is an imprint of Boom uitgevers Den Haag.

ISBN 978-94-6236-810-1
ISBN 978-94-6274-800-2 (E-book)

Printed in The Netherlands

Table of Contents

Abbreviations

AC	alternating current
ACEF	Asian Clean Energy Fund (CEFPF)
ADB	Asian Development Bank
AEB	Afval Energie Bedrijf (Waste to Energy Company, Amsterdam)
AEC	Afval Energie Centrale (Waste-to-Energy Plant, AEB)
AfDB	African Development Bank
AG	Advocate General (of the CJEU)
AHP	analytic hierarchy process
APP	advanced plasma power
ASR	automotive shredder residue
BAT	best available technology
BFB	bubbling fluidized bed
BOT	build–operate–transfer
BTEX	pyrene (benzene, toluene, ethylbenzene and xylenes)
C	celsius
CARICOM	Caribbean Community
CC	combined cycle
CCHP	combined cold, heat and power
CCP	combined cold and power
CCSF	Carbon Capture and Storage Fund (CEFPF)
CCU	Carbon Capture Utilisation
CCUS	Carbon Capture Utilisation and Storage
CDM	Clean Development Mechanism (Kyoto)
CE	circular economy
CEF	Clean Energy Fund (CEFPF)
CEFPF	Clean Energy Financing Partnership Facility (ADB)
CERs	Certified Emission Reductions (CDM)
CEWEP	Confederation of European Waste-to-Energy Plants
CFB	circulating fluidized bed
CFC	chlorofluorocarbon
CFPS	Canadian Climate Fund for the Private Sector in Asia (CEFPF)
CH_4	methane
CHP	combined heat and power
CI	compression ignition
CIFs	Climate Investment Funds
C&IW	commercial and industrial waste
CJEU	Court of Justice of the European Union
Cl	chlorine
CMW	combustible materials derived from waste
CO	carbon monoxide
CO_2	carbon dioxide
CR	consistency ratio coefficient
CTF	Clean Technology Fund (CIF)
DBO	design–build–operate
DC	direct current

DG	Directorate General (European Commission)
DG CLIMA	Directorate General for Climate Action (EU)
DG ENER	Directorate General for Energy (EU)
DG ENV	Directorate General for the Environment (EU)
DMS	direct melting system
EAP	Environmental Action Programme
EBRD	European Bank for Reconstruction and Development
EC	European Community
ECAFE	Economic Commission for Asia and the Far East (UN)
ECJ	European Court of Justice
Eco-AP	Eco-Innovation Action Plan (EU)
ECR	European Court Reports
EEA	European Environment Agency
EFSI	European Fund for Strategic Investment
EHS	Environmental Health and Safety (Guidelines-IFC)
EIA	Environmental Impact Assessment
EIB	European Investment Bank
ELFM	enhanced landfill mining
ELV	end-of-life vehicles
ERF	energy recovery facility
ESP	electrostatic precipitator
ETS	EU Emissions Trading Scheme
EU	European Union
EUP	pressurized gasification and slagging process
EWM	enhanced waste management
FIP	Forest Investment Program (SCF)
GCC	Gulf Cooperation Council
GEF	Global Environment Facility (WB)
GHG	greenhouse gas emissions
GPP	Green Public Procurement
GT	gas turbine
GW	gigawatt
H	hydrogen
HCL	hydrogen chloride
HFC	hydrofluorocarbon
HRC	Hoog Rendement Centrale (High Efficiency Waste Fired Power Plant, AEB)
IBRD	International Bank for Reconstruction and Development (WB)
ICE	internal combustion engine
IDA	International Development Agency (WB)
IDB	Inter-American Development Bank
IE	industrial emissions (Directive-EU)
IET	International Emission Trading (Kyoto)
IFC	International Finance Corporation (WB)
IFC PS	International Finance Corporation Performance Standards
IFEU	Institute for Environmental and Energy Research
IGCC	integrated gasification combined cycle
IGFC	integrated gasification fuel cell
IPP	Integrated Product Policy (EU)
JI	Joint Implementation (Kyoto)
JRC	Joint Research Center
KOEL	Karachi Organic Energy Limited

LCA	life cycle assessment
LFD	Landfill Directive (EU)
LFG	landfill-gas
LHV	lower heating value
LIFE	L'Instrument Financier pour l'Environnement (EU)
MBT	Mechanical Biological Treatment
MCA	multi-criteria analysis
MJ	megajoule
MSW	municipal solid waste
MW	megawatt
N	nitrogen
NABU	Nature and Conservation Union
NGO	non-governmental organization
NH_3	ammonia
NO_x	nitrogen oxides
O	oxygen
OECD	Organisation for Economic Co-operation and Development
OGEL	Oil, Gas & Energy Law Intelligence
OJ	Official Journal (EU)
OJEU	Official Journal of the European Union
OPEX	Operational Expenditure
PAH	polyaromatic hydrocarbons
PCBs	polychlorinated biphenyls
PCDD	polychlorinated dibenzodioxins
PCDF	polychlorinated dibenzofurans
PCF	Prototype Carbon Fund (WB)
PF4EE	Private Finance for Energy Efficiency (EIB)
PJ	petajoules
PPB	parts per billion
PPCR	Pilot Program for Climate Resilience (SCF)
PPM	parts per million
PPMV	parts per million volume
PPP	public–private partnership
RDF	refuse-derived fuel
RED	Renewable Energy Directive (EU)
RRR	reuse, recovery and recycling
RT	residence time
S	sulphur
SCF	Strategic Climate Fund (CIF)
SCP	sustainable consumption and production (EU)
SI	spark-ignition
S-LCA	social life cycle assessment
SMM	sustainable materials management
SO_2	sulphur dioxide
SOFC	solid oxide fuel cell
SREP	Scaling-Up Renewable Energy Program in Low Income Countries (SCF)
SRF	solid-recovered fuel/specified recovered fuel
TFEU	Treaty on the Functioning of the European Union
UBA	German Federal Environment Agency
UNEP	United Nations Environment Programme
UNESCAP	United Nations Economic and Social Commission for Asia and the Pacific

UNFCCC	United Nations Framework Convention on Climate Change
UNTS	United Nations Treaty Series
VOCs	volatile organic compounds
WB	World Bank
WEEE	waste of electrical and electronic equipment
WFD	Waste Framework Directive (EU)
WtE	waste to energy
WtM	waste to material
WtP	waste to products
WUR	Wageningen University & Research

Preface

The European Union purports to be moving towards a circular economy (CE). If recent experience in environmental and energy law is anything to go by, the rest of the world will look with interest to its progress. It is fashionable to say that in the CE, "waste" will no longer exist. This is however not relevant beyond semantics. What really matters is how the EU and others, after or before it, create the economic and regulatory environment that enables the innovation that a CE requires.

Regulatory circles have ample sympathy for business implementing and bringing to market the many exciting ideas that engineers continue to develop. At the same time, one must not be blind to the excesses that unchecked engineering imagination might have on society, in all pillars of sustainable development: social, economic and environmental. We must not compromise on a robust regulatory framework that looks after what public health and environmental protection require: Late Lessons from Early Warnings reports tell us that we would do so at our own peril. However, we do have to question continuously whether our existing laws are best practice in reaching that desired outcome. It would be a particular affront if innovative products and services that truly may boost environmental protection, were not to be rolled out because of anxiety over their legal status.

In an innovative environment, legal certainty is an important driver for success. Lack of clarity over the legal framework, and/or the regulators' implementation of the same, leads industry either to seek out and concentrate development on those states with lax or flexible regulators only or to stick to old and trusted products.

The European Union is particularly suited to providing that clarity: On the scientific front, by investing in research and development, especially at the level of small or medium-sized enterprises (SME) and at the specialized spin-off level. On the regulatory front, by working out a regime that enables innovators to query enforcement agencies about the legality of a new product or service line without the fear of subsequently being disciplined for it.

This volume is a scholarly effort to assist with both strands of the exercise. It is to be much commended for that effort and I for one am sure both industry and legal scholars will find its content encouraging.

Geert van Calster
Leuven, September 2017

List of Contributors

Sietse A. Agema	Amsterdam Energy Company (AEB), Strategic Advisor
Nicholas Agon	KU Leuven, Department of Mechanical Engineering, Postdoctoral Research Fellow
Anouk Bosmans	Chemical Engineer (MSc KU Leuven)
Filippo Corsini	Sant'Anna School of Advanced Studies, Pisa, Institute of Management, Research Fellow
Thomas J. de Römph	PhD Candidate at the Law Faculties of KU Leuven and Hasselt University
Marco Frey	Sant'Anna School of Advanced Studies, Pisa, Institute of Management, Professor of Economics and Business Management; Research Director of the Centre for Research on Energy and Environmental Economics and Policy (IEFE), Bocconi University, Milan
Natalia Gusmerotti	Sant'Anna School of Advanced Studies, Pisa, Institute of Management, Research Fellow
Lieve Helsen	KU Leuven, Department of Mechanical Engineering, Professor of Mechanical Engineering
Harry H.G. Post	Faculté libre de droit, Université Catholique de Lille, Adjunct Professor; Dipartimento di Giurisprudenza, Università degli Studi di Sassari, Visiting Professor; University of Exeter, Professor of International Law, ret.
Stefano Silingardi	Dipartimento di Giurisprudenza, Università degli Studi di Modena e Reggio Emilia, Assistant Professor
Kim Talus	James McCulloch Chair in Energy Law, Tulane Law School and founding Director of the Tulane Center for Energy Law, New Orleans, United States; Professor of European Economic and Energy Law, Law School, University of Eastern Finland & Professor of Energy Law, University of Helsinki, Finland
Topi Turunen	University of Eastern Finland, Law School, and at the Finnish Environment Institute, Doctoral Researcher
Geert Van Calster	KU Leuven, Head of the Department of International and European Law; Professor of Law
Matthias Wieseler	AEB, Plant Technology Team leader

Introduction

This book is the third volume of the series, which is the product of an informal research project entitled *Waste in International and European Law and Policy: Tension between Trade and Environmental Protection, between Union and Member State*. This somewhat unusual research project has benefitted from sufficient (financial) support from a diversity of sources and in particular from a score of interested researchers since it was started in 2012. Besides some smaller articles and notes, it has so far produced three edited volumes on waste management: all three were inspired by International workshops that took place at the Faculté Libre de Droit of the Université Catholique de Lille and at the Università degli Studi di Sassari, Italy (where it was supported by various departments). The first of these books is entitled *Waste Management in European Law, the Example of Naples and Campania*; the second *Radioactive Waste Management in International and European Legal Perspective*. Whereas in the first two books, as their titles say, a legal perspective prevailed, in this third volume a "really" multidisciplinary approach has been adopted: economists, social scientists, lawyers and engineers have worked here together. "Waste to energy" is a subject matter that is very much in development. Important economic and policy choices still have to be made, in Europe and elsewhere. Accordingly, its legal regulation is often still rather rudimentary, and changing. Its future will undoubtedly strongly depend on the contribution it can make to a circular economy and its energy transition. In that perspective, the future of waste-to-energy technologies seems to look promising enough as the investments already materialized and expected, show. However, quite a number of factors and conditions are as yet uncertain as is amply demonstrated in the contributions to this book. Even the technological "foundation" of waste to energy is a subject matter that still remains to be seriously discussed as the pages before you show.

In an assessment of the "state of affairs" of waste-to-energy processes as a way of managing waste and as a source of energy, the first chapter probes the history and public image of waste to energy (WtE). From a technical point of view, the WtE processes have improved very substantially since incineration of waste caused major problems for public health and the environment. The regulation of WtE processes followed suit (in particular, ever more sophisticated environmental impact assessment (EIA) regulation, as did information duties and greater transparency of decision-making. WtE plants increasingly came to be seen and treated as "ordinary" middle-sized industrial plants, both in respect to the regulation of their emissions and in terms of their location. Gradually, it may be assumed – lacking sophisticated analysis – that the public image of WtE also improved. In fact, in modern times, it seems that WtE even enjoys considerable social acceptance as a result of its role in transforming municipal solid waste (MSW) into locally/municipally used electricity or public heating. The European Commission is even somewhat concerned about the support of some Member States of WtE. European Commission activity with regard to the sector has perhaps not been overwhelming, even somewhat ambivalent, Post submits in this chapter. There is clearly room and even a need for further elucidation of the place and development of WtE processes in the future, including within the circular economy strategy perspective. The uneven availability of WtE plants in the European Union and the future availability of solid waste fuel to feed those plants is a subject of serious concern, including of course for the WtE sector itself. The resulting and potentially increasing market of solid waste leading to growing flows of such waste in Europe as well as from other parts of the world is a development requiring a more adequate assessment and regulation.

In Chapter 2, Stefano Silingardi analyses the current investment policies in WtE. He provides data showing a doubling of global waste generation and a compound annual growth rate of the WtE market of 5.5% from 2016 to 2024 to a size of almost US$44 billion. Although debt and equity still maintain their leadership as the most effective investment option for WtE infrastructure projects in the developed world, he observes that the picture in the developing world is rather different. Here solid waste management generally has a very low priority. Very limited funds are available and, even more problematic, the local taxation system is usually inadequate and, as services are irregular, users are not willing to pay either. For such countries, the author argues WtE has proved to be attractive as it also contributes to the resolution of energy needs in a situation of vigorous urbanization where the generation of MSW is (very) high. In view of the traditional difficulty for developing countries to attract foreign capital, international actors like multilateral investment banks (i.e. the World Bank group, the Asian Development Bank, or the ERDB, the European Bank for Reconstruction and Development) are increasingly getting involved in that sector. In addition, new initiatives have developed to stimulate initiatives that increase the use of WtE processes in order to facilitate the energy transition to low-carbon economies.

In Chapter 3, Lieve Helsen, Nicholas Agon and Anouk Bosmans review and compare incineration, pyrolysis, gasification and plasma gasification – the four most important thermochemical treatment technologies for MSW and/or refuse-derived fuel (RDF). Most WtE processes require a pre-processed form of MSW as the chemistry and fluid dynamics of more advanced technologies are very sensitive to feedstock variations. For all these technologies, a more homogeneous waste input improves operating conditions as it allows tighter process control, and thus more stringent product quality specifications can be met. The authors present a detailed overview of the four thermochemical processes in waste management they focus on. In respect to *incineration*, they emphasize the large potential of heat/electricity this process is able to offer without extensive pre-treatment, in particular MSW incinerators where combined heat and power (CHP) is applied. The use of steam-turbines does not guarantee optimal flue-gas quality, but modern WtE incineration plants can reach high net electricity efficiencies. However, although such solid waste incinerators can reduce the waste volume to about 90%, they require advanced thermal treatment options in order to avoid that still substantial volumes of residue and the bottom ash end up in landfills and to make the fly-ash safe. With regard to *pyrolysis*, the necessity of substantial waste pre-treatment and/or after-treatment constitutes the main problem. Although it has proved its use for specific feeds like tyres or electronic waste, the potential presence of pollutants could make the resulting products useless. The authors conclude that the stringent conditions for this process make it less suitable for the rather heterogeneous MSW and are likely to bar its commercial viability. *Gasification* has several advantages over traditional combustion of MSW: the formation of dioxins, SO_2 and NO_x is limited and the volume of process gas is low, requiring smaller and less expensive gas cleaning equipment. Moreover, the volume reduction of the waste is more than in the case of incineration. The generation of combustible gas allows more efficient electricity production than steam turbines. But due to economic and practical reasons, the syngas produced is still often burned in steam boilers. Like, for *pyrolysis* the highly heterogeneous nature of MSW causes significant challenges with respect to the gasification process and feedstock handling. More advanced processes might offer a better perspective. The authors find *plasma gasification* to be the most promising of the processes they have dealt with. However, it is a relative newcomer in the field of MSW/RDF treatment. Plasma technologies do not use thermal energy freed from combustion, but use electricity as the source of heat, a source that is thus independent from the heated substances. This leads to more flexibility of the system and fast process control. Moreover, it provides more options in process chemistry including the possibility of generating more valuable substances out of waste. So far, due to its high costs, as electricity is still relatively expensive, plasma technology has been used mainly for the treatment of hazardous waste. As a result of improved design and the added value of the production of more marketable products, plasma systems have the potential of widening their foothold in waste management. The authors argue that

> [f]rom an environmental point of view, plasma systems have the potential to offer a viable alternative to landfilling and conventional incineration, with lower air pollution and virtually no residual waste streams requiring final disposal.

Provided that regulatory, economic and sociopolitical drivers continue to encourage the implementation of advanced thermochemical conversion technologies, they expect that the commercial viability of plasma technologies will improve in the future.

Chapter 4 focuses on the application of WtE processes in one plant, a particularly successful application one may say: the "Amsterdam Energie Bedrijf", the Amsterdam Energy Company or AEB. Following a review of the over a century old efforts of the municipality of Amsterdam, the Netherlands, to cope with its waste, primarily dealing with it as a health hazard, the authors argue that now the approach has changed. After major environmental disasters with the municipal incinerator in the 1980s, the next-generation WtE plant was successfully designed to be clean. The current WtE plant has become foremost, as the authors say, "…'*designed for output*'". The plant maximizes the recovery of energy and materials. The authors point to the record 30% net electrical efficiency the AEB technology produces, and emphasize that gasification is still not proven to be an equally robust technology for untreated MSW. In a circular economy, WtE should eventually focus on treating materials that cannot be cost-effectively recycled due to their composition or way of use. As MSW consists of over 50% of materials with biomass origin, it is a sustainable renewable energy source. The replacement of electricity production from an energy mix of over 50% renewable source means a huge CO_2 reduction. Besides energy recovery, material recovery as in metals and non-metals from the slacks (burning residue) or inert fraction of the MSW, also adds to a very considerable reduction of CO_2 emissions compared to landfilling. They proceed then with a detailed explanation of the high-efficiency technology developed and applied by the AEB explaining its dependability (an availability of over 90%) and stability over the years: it must now be regarded as proven technology, making it a clean and safe sink solution for a circular economy. In conclusion, they submit that at least in a transition period – which may last some decades – to a (more) circular economy, thermal conversion of residual waste streams using high-efficiency technology will be necessary. Whenever possible, the location of this type of WtE plant should be in the vicinity of other process industries, chemical industries and the cities' sewage plants in order to gain maximum industrial symbiosis.

In Chapter 5, Filippo Corsini, Natalia Gusmerotti and Marco Frey report on an analytic hierarchy process (AHP) assessment of the feasibility of four WTE technologies as used in the United Kingdom, Denmark and Italy. By using economic indicators, energy recovery efficiency, environmental impact and suitability with respect to waste streams and the separate collection system, they assess co-combustion in a coal-fired power plant, waste incineration with energy recovery, gasification and biogas production from anaerobic digestion in order to measure their circularity. The study shows that in each of these countries the most feasible WtE technology is different. AHP will help to increase knowledge about the feasibility of WTE technologies in a circular economy perspective, they submit, and thus assist decision-makers and technicians to fully exploit the potential for energy recovery of non-recyclable waste. Comparing the situation in several countries as has been done in this chapter will also bring further clarity about the position of energy recovery in the waste hierarchy and the achievement of the circular economy objectives. Adding further criteria, in particular, job creation, and using primary data instead of data found in the literature, will make it possible to understand the underlying dynamics even better.

Using waste as a resource to generate energy would save fossil fuels used in conventional power plants. It would contribute both to energy security and to getting rid of waste and diverting it from landfills. In addition, materials can be extracted from waste thus saving or replacing virgin materials. Moreover, both using the energy in waste and using the materials extracted from waste would mean lower CO_2 and other greenhouse gas emissions than in the case of the use of the virgin equivalents. In Chapter 6 Thomas De Römph discusses the various policies the European Union has issued concerning the triangular nature of the material–waste–energy cycle and the opportunities WtE technologies offer here. The most influential and important are the two-pronged policies in respect to energy and waste. In 2015, a new "encompassing" approach was launched: the "Circular Economy Package" (or CE package). Establishing an adequate regulatory framework to enable the transition by 2050 to a circular economy is one of the crucial challenges ahead for the EU. The chapter focuses on the consequences of the circular economy plans for WtE practices. Following an overview of the existing regulation of WtE, De Römph continues by discussing in detail the way the CE proposals would change this framework. As a general rule, he argues, the EU Waste Framework Directive "...favours Waste-to-Materials over Waste-to-Energy in the waste hierarchy, because it favours preparing for reuse and recycling over other recovery options and disposal activities respectively." Based on existing EU waste law, the circular economy transition is above all a quest to transform the materials system, amongst others through law. It means that EU materials law, which not only addresses the waste stage but also other life cycle stages, will be adapted and strengthened in view of this goal. In that light, "waste to energy" is regarded as a positive side effect of the circular economy and in that way as merely helping the energy transition.

In the final chapter, Chapter 7, Kim Talus and Topi Turunen discuss the danger of bad "old" legislation by way of three examples. What the three legislations have in common is that they each pose barriers to innovation in WtE technology, notably to the manufacturing and use of "producer gas" (i.e. gas produced from another product instead of being obtained from natural resources). Under the current regulation, producer gas is subjected to a double regulatory threshold. On the one hand, it is subject to the normal end-of-waste regulation. On the other hand, its emissions are compared to natural gas emissions and accordingly evaluated. Its use in (advanced) WtE plants is much more efficient than the use of natural gas and can lead to an increase in electricity production of over 60%. The analysis in this chapter focuses on the EU Industrial Emissions Directive (IE-Directive), which provides the rules on integrated prevention and control of pollution from industrial activities, and, notably, on the new Article 42 (1). This article was added following the (2008) Judgment of the European Court of Justice in *Lahti Energia Oy*. Both the Court and the Advocate-General in that case suggested that producer gas could have the same qualities as fossil fuel. Talus and Torunen's analysis shows, however, that Article 42 (1) is inadequate. The provision requires that producer gas is compared with natural gas, which does not lead to treatment on equal footing of the two. However, they argue, whereas these two gasses differ in composition, they should be treated distinctly in the relevant regulation (like the IE-Directive). The current regulation, they argue, poses difficulties – and they continue to demonstrate in detail which ones – both for the licensing authorities and for potential investors. Such regulatory hurdles should be eliminated so that the market can pick the best technology to attain a particular objective. Such regulation should also include the stipulation of correct emission values so that the environment is well protected.

A major source of inspiration for the book – producing critical questions, providing new ideas and opening new venues to approach – was an International Workshop/Journée d'Etude Internationale entitled *From Waste to Energy: Technology, the Environment and the Implications under EU Law*, held in Lille (France) in October 2015. It was hosted and supported by the Faculté Libre de Droit, Université Catholique de Lille, and was generally considered a successful meeting primarily as a result of the most interesting contributions of the expert speakers but also because of the participation of a large and engaged audience. Dr Ioannis Panoussis and Dr Aurelien Raccah, respectively dean and vice-dean of the Faculty, assured generous hospitality and they fully deserve gratitude. So does the staff of Eleven International Publishing for their most competent help and their patient support through the whole process from manuscript to book.

Harry H.G. Post
Bologna-Sassari, September 2017

1 From Waste to Energy: The State of Affairs (with a Focus on the European Union)*

Harry H.G. Post

Although research in the development of the most adequate waste to energy (WtE) technology and its most practical application is still an ongoing challenging task,[1] investment in WtE plants is already rather substantial and increasing. In the first years of this decade, venture capital and private equity business investment in the WtE sector have been estimated to have increased by 186% to a total of US$1 billion. The growth is exponential. According to the World Energy Council, the WtE market as a whole is expected to reach close to US$30 billion by 2015 and almost US$44 billion by 2024 with major expansions in China and India as well as in the European Union.[2] In the EU, cumulative investment in municipal waste management over the 10 year period 2011-2020 is estimated at US$87 billion.[3]

It seems then that municipal solid waste (MSW) may have, indeed, "…the opportunity to become a precious source and fuel for the urban sustainable energy mix of tomorrow."[4] Unlike other energy-producing operations, WtE plants have two purposes: generation of energy and management of waste, notably MSW. They do the latter by performing their traditional function of reducing the waste volume considerably, leaving less eventually to deposit in a landfill. In addition, they produce, and increasingly do so, useful materials in the process. Energy production and solid waste management are both complicated processes, which have a substantial environmental impact. Some 40 years ago, management of solid waste by way of "plane" incineration had acquired a rather bad reputation. Although waste incineration for a long time already was the prominent management technique to reduce the amount of solid waste to deposit, it had widely come to be seen as a substantial public health hazard. In 1971, W. M. Harrington opened a pre-

* I wish to thank Ige Dekker and Paul Lucardie for their valuable comments.

1 *See* the extensive account by Stefano Silingardi in Chapter 3 of this book.

2 World Energy Council, World Energy Resources-2013 Survey, London 2013, ch. 7 b. 2. The estimated market size of the sector in 2015, however, is currently taken to be US$28.86 billion (*see, also* for the 2024 estimate: <www.statista.com/statistics/480452/market-value-of-waste-to-energy-globally-projection/>; *see*, further ch. 3, below).

3 Colsweep, *Report: Global Analysis of the Waste-to-Energy field*, 2013, p. 23. *See also* CEWEP, *Waste-to-Energy Plants in Europe in 2014*, March 2016, available at: <www.cewep.eu/ information/data/studies/ m_1488>.

4 World Energy Council, *op. cit.* footnote 2, Ch. 7 b. 2.

sentation on waste incineration before the American Public Works Association in the following way:

> There are probably no more controversial subjects today than air pollution and incineration. After more than 100 years of relatively successful operation and general acceptance as a volume reduction process, incineration has now become unacceptable to many environmentally-conscious individuals.[5]

He continued his presentation by trying to convince his audience that as a whole the problems in waste incineration were not as severe as they might seem in the public eye of his time. Still nobody, including Harrington, denied that the operation of many incineration plants seriously polluted their environment. The biggest concern at the time was the production of a huge amount of furans and dioxins.[6] In his concluding remarks to the Association, Harrington nevertheless pointed out a great benefit of waste incineration – in his time primarily attempted in Europe – in his argument:

> Incineration has the potential to provide resource saving through utilization of the refuse heat content and resource recovery through recycling of the valuable components of the ash residue.[7]

This potential of incineration could be fully realized, he concluded, if reasonable emission regulations, gradually increasing in stringency, would accompany improvements in incinerator design and air pollution control technology.

Incinerator design and air pollution control technology have certainly improved very considerably since the 1970s and 1980s and research towards further improvement has not stopped ever since. Better design and improvement in air pollution control technology and its application, gradually have resulted in what amounts to virtually a stop of any significant air pollution from MSW incinerating plants. The EU Directive on Industrial Emissions (previously "Integrated Pollution and Prevention Control") can now realistically set severe conditions for the operation of waste incineration and co-incineration plants.[8] Annex VI on the "Technical provisions relating to waste incineration plants and co-incineration plants" addresses in detail the prevention of pollution from SO_2,

5 W.M. Harrington, Jr, 'Incineration and the Air Pollution Problem', Presentation at the APWA Public Works Congress and Equipment Show, September 1971, p. 21 (reprinted at <www.seas.columbia.edu/earth/wtert/sofos/nawtec/Incinerator-and-Solid-Waste-Technology/Incinerator-and-Solid-Waste-Technology-04.pdf>).

6 *See also* Chapter 4 by Agema and Wieseler in this book.

7 *See op. cit.*, footnote 5, p. 24.

8 Defined in Article 3 under 40 and 41, and addressed in its Annex I to Directive 2010/75/EU of 24 November 2010, on industrial emissions (integrated pollution prevention and control); OJEU of 17.12.2010, at L 334/17 ff.

NO_x, dust, CO, etc. including, in its Parts 3 and 4, the emission of dioxins and furans. Emission controlling bodies all over the world can now require incinerators to eliminate more than 99% of the toxins present during waste processing whereas ash residue left in the combustion chamber generally is supposed to be taken to a hazardous-waste landfill.[9]

But apart from the need to improve the technology of incinerators and the regulation of emissions, the "image" of incineration, including WtE incineration, no doubt needed some improvement as well. The most formidable barrier against incineration that had arisen was not technological even if the benefit of providing electricity and heating to a nearby urban population was taken into account. A famous 1985 study argued that public opposition had become the greatest single obstacle to the successful siting of WtE projects in California: people tended to view WtE projects in the same light as any waste disposal facility, and they simply did not want a "dump" nearby.

The Californian study tried to find out who in a community objected most and why, and what could be done to improve the situation. Extensive research based on questionnaires found that public concern is often rooted in previous experiences with poorly managed waste facilities, especially hazardous waste landfills. Such viewpoints, it argues, may be difficult to overcome even by reasonable arguments based on undeniable advances of modern technology and its beneficial side effects. The study concludes that a concerted public participation programme is needed. It should be justifiably wrapped in some of the very pro-environmentalist arguments used against landfills. Then such a programme can be a highly effective tool in educating the public as to the true nature of WtE projects.[10] An attempt at assessing the state of the "public image" of WtE technology and plants in the light of some of the relevant regulations will be at the centre stage in the following sections.

1.1 Environmental Impact Assessment

By the end of the last century, the image problems with incineration plants can be said to have been understood "in full", which is quite something else than addressing them adequately. A 1999 World Bank Technical Guidance Report, e.g. states as follows:

---

9 *See*, on dioxins and furans: "An Investigation into the Performance (Environmental and Health) of Waste to Energy Technologies Internationally; Summary Report – Waste to Energy – A review of legislative and regulatory frameworks, state of the art technologies and research on health and environmental impacts," Compiled by WSP Environmental for the Government of Western Australia Department of Environment and Conservation, 2013, pp. 32/3; and an overview of emission requirements in several EU countries, Japan, Australia and the United States at pp. 7-19.

10 'Political Difficulties Facing Waste-to-Energy Conversion Plant Siting', Cerrell Associates, Inc., 1984 Study for the California Waste Management Board, p. 40.

> MSW incineration plants are generally located close to densely populated areas for economic reasons (e.g., sale of energy, transport distance). Any negative environmental effects of the plant can influence a great number of people. A combination of planning and technical measures is required to minimize such impacts.[11]

The World Bank Report continues by paying particular attention to the negative environmental effects of a waste incineration plant, which is considered comparable to medium to heavy industrial activities as far as its environmental impact is concerned. A full environmental assessment is required for such plants. This has become the standard requirement in practice. In the 2011 EC Directive on Environmental Impact Assessment, for instance, Waste disposal installations for incineration or chemical treatment as defined in Annex I to Directive 2008/98/EC (i.e. the Waste "Framework" Directive) under heading D9 of non-hazardous waste with a capacity exceeding 100 tonnes per day are subject to a full-scale environmental impact assessment in accordance with Articles 5-10 of the EIA Directive.[12] In the EU countries as well as in many places elsewhere, authorities in principle issue a permit to operate an incinerator only when the environmental impact assessment (EIA) is satisfactory.[13] It depends somewhat on the nature of the instrument and the precise factors that are concerned, but by means of an EIA always a great variety of them must be assessed.

Under the broad UNEP (United Nations Environmental Program) definition, environmental impact assessment is "a tool used to identify the environmental, social and economic impacts of a project prior to decision-making." An EIA may not cover all kinds of other location factors including the nature of the process (i.e. the way energy is produced) or the availability or quality of the waste. Such aspects, like the importance of the attitude of the local population, may be dealt with in other (legal) instruments specifically addressing waste management. Usually, however, the EIA of a project soon takes centre stage over other considerations in particular if its results are public and its requirements range widely. The EIA requirements are laid down in numerous national laws and in various international instruments. Per installation the specific EIA requirements can therefor differ greatly in substance and in legal nature but they can also differ nationally and locally to the extent that it is difficult to provide a general picture for specific types of plants like WtE installations. If more complicated assessment techniques are used, like in the so-called life cycle analysis (LCA) approach for products and services, differences per

---

11 'Municipal Solid Waste Incineration; a World Bank Technical Guidance Report', The International Bank for Reconstruction and Development/World Bank, Washington, p. 87.

12 Annex 1, under 10, of Directive 2011/92/EU of 13 Dec. 2011 on the assessment of the effects of certain public and private projects on the environment (in OJEU, L 26/1 ff. of 28/1/2012).

13 The following pages are liberally based on my 'Waste to Energy Plants: Location and Popular Protest in the EU (in a Legal Perspective)'; in 3 OGEL, 2016 (<https://www.ogel.org/article.asp?key=3625>).

location of the project under consideration will increase even further, leading easily to the (preliminary) conclusion that only a case-by-case evaluation will lead to acceptable insights into the effects of major waste disposal plants. This will be even more so if LCA is used and is explicitly extended to include the consequences for people (the social life cycle analysis, or S-LCA).[14]

For the European Union, EIA focuses strongly on the impact on the environment. Article 2 of the detailed EU "codification" Directive 2011/92/EU on the assessment of the effects of certain public and private projects on the environment, requires that

> Member States shall adopt all measures necessary to ensure that, before consent is given, projects likely to have significant effects on the environment by virtue, inter alia, of their nature, size or location are made subject to a requirement for development consent and an assessment with regard to their effects.[15]

Article 3 of the EIA Directive prescribes that the assessment concerns the effects of a project on

> (a) human beings, fauna and flora; (b) soil, water, air, climate and the landscape; (c) material assets and the cultural heritage; (d) the interaction between the factors referred to in points (a), (b) and (c).

Although in an EIA in the EU attention is thus also to be paid to the effect of a project on "human beings", as yet this provision probably did not have much of a positive effect on the public image of waste incineration. At best, it may perhaps have had an impact on how authorities deal with local protest against the establishment of a waste disposal installation. What may have helped more is that gradually, even in initial siting phases, information on a project and often consultation of the people concerned, have become part of standard procedures.

This is in particular supposed to be so since the EU has become a party to the 1998 Aarhus Convention on Access to Information, Public Participation in Decision-making and Access to Justice in Environmental matters.[16] Since 2003, the EU has been actively amending relevant legislation like the EIA Directive 2011/92/EU according to "Aarhus-type" principles.[17] As a result, in principle, public information and participation in rele-

14 A social and socio-economic Life Cycle Assessment (S-LCA) is a social impact (and potential impact) assessment technique that also aims to assess the social and socio-economic aspects. Cf. UNEP, 2010 Guidelines for Social Life Cycle Assessment of Products, 2009, p. 37.

15 Directive 2011/92/EU *op. cit.*, footnote 12.

16 [1999] 2161 UNTS 447, or 38 International Legal Materials (1999), p. 517 (in force since 29 October 2001, for the EU since May 2005).

17 The 2003 amendments to the EIA Directive 2011/92/EU (*op. cit.*, footnote 12) or, *see* Annex IV to Directive 2010/75/EU on industrial emissions (integrated pollution prevention and control), OJ 2010 L 334/17. The

vant decision-making procedures is now supposed to be improved very considerably and certainly in comparison to the situation in the 1970s and 1980s. We may presume that following the (very) considerable technological progress, such improvement on transparency in respect to the siting and operation of WtE plants, also has a positive effect on its public image. This "image improving" effect of the Aarhus modifications of course might only be supposed if the application of these principles in fact takes place in practice. It does appear that the Aarhus process towards greater transparency is (at least) still rather cumbersome, but definitely somewhat controversial including the way the Court interprets it.[18] Initially the developments looked promising enough as, e.g. the General Court rejected two decisions of the European Commission with a restrictive approach to the so-called Aarhus Regulation.[19] The Court did so by referring directly to Article 9(3) of the Aarhus Convention. But, on appeal, the Court of Justice of the EU (CJEU) rejected these arguments and avoided a legal review of the Aarhus Regulation in this light.[20] Where this leaves us is not entirely clear. Still, following a detailed analysis of the developments in the ECJ jurisprudence, Jans and Vedder conclude that "Article 9(3) of the Aarhus Convention is applicable across the full breadth of European environmental law." They also hold that now "...environmental groups should be allowed access to a court to challenge decisions that might conflict with the environmental law of the Union."[21] We may assume

---

basis for these amendments were two 2003 Directives, on, respectively, Public Access to Environmental Information (2003/4/EC of 28 January 2003) and on Public Participation (2003/35/EC, of 26 May 2003.

18 *See, e.g.* Hendrik Schoukens, 'Access to Justice in Environmental Cases after the Rulings of the Court of Justice of 13 January 2015: Kafka Revisited?', Utrecht Journal of International and European Law, Vol. 31, No. 81, 2015.

19 Council Regulation (EC) No 1367/2006 of 6 September 2006 on the Application of the Provisions of the Aarhus Convention on Access to Information, Public Participation in Decision-making and Access to Justice in Environmental Matters to Community institutions and bodies [2006] OJ L 264/13 (Aarhus Regulation, into application since 17 July 2007).

20 Cases T-396/09 *Stichting Natuur en Milieu and Pesticide Action Network Europe v. Commission* (CFI, 14 June 2012), and T-338/08 *Vereniging Milieudefensie and Stichting Stop Luchtverontreiniging Utrecht v. Commission* (CFI, 14 June 2012). On appeal: Joined cases C-404/12 P and C-405/12 P *Council and Others v. Stichting Natuur en Milieu and Milieu and Pesticide Action Network Europe* (CJEU, 13 January 2015) and Joined cases C-401/12 P and C-403/12 P *Council and Others v. Vereniging Milieudefensie and Stichting Stop Luchtverontreiniging Utrecht* (CJEU, 13 January 2015).

21 Jan H. Jans and Hans H.B. Vedder, European Environmental Law; after Lisbon, 4th ed., Groningen Europa Law Publishing, 2012, p. 237. The text of Art. 9.3 says: "In addition and without prejudice to the review procedures referred to in paragraphs 1 and 2 above, each Party shall ensure that, where they meet the criteria, if any, laid down in its national law, members of the public have access to administrative or judicial procedures to challenge acts and omissions by private persons and public authorities which contravene provisions of its national law relating to the environment." Jans and Vedder's conclusion seems (tentatively) warranted by the recent (28 April 2017) Notice on Access to Justice in Environmental Matters clarifying how individuals and their associations can challenge decisions, acts and omissions by public authorities related to EU environmental law before national courts (EC Communication C (2017) 2616).

then that all in all also the Aarhus Convention, together with other improvements in the applicable EIA instruments, contributes to a better public image of WtE technology.

1.2 EU Waste Legislation

Besides the general EIA regulations, more specific modern waste management legislation may have further improved WtE's public image by setting additional conditions to waste management and WtE plants. WtE incineration plants concern non-hazardous urban waste. Like all waste incinerator plants, they are also subject to location criteria, one of which is the environmental impact assessment we have just discussed. In addition, Article 13 on "Protection of Human Health and the Environment" of Directive 2008/98/EC, the Waste "Framework" Directive, requires that:

> Member States shall take the necessary measures to ensure that waste management is carried out without endangering human health, without harming the environment and, in particular:
> (a) without risk to water, air, soil, plants or animals;
> (b) without causing a nuisance through noise or odours; and
> (c) without adversely affecting the countryside or places of special interest.

This article, under (b), can be said to concern a location criterion of some relevancy in view of our subject matter: the continuous transport through a residential area of large quantities of urban waste may, e.g. amount to levels of noise or odours that could harm the environment, or the enjoyment of homes and privacy, without necessarily endangering human health. Hence, opposition of local residents to waste transport based on such nuisance may be considered a valid objection, preventing a permit for a waste disposal installation to be issued by the competent authority.[22] However, Article 23(3) of the 2008 Waste Framework Directive on "The issue of permits" provides only one environmental ground for refusing a permit:

> [w]here the competent authority considers that the intended method of treatment is unacceptable from the point of view of environmental protection, in particular when the method is not in accordance with Article 13, it shall refuse to issue the permit.

22 Article 23.1 of Directive 2008/98/EC of the European Parliament and of the Council of 19 November 2008 on waste and repealing certain Directives (OJEU L 312/3 ff of 22 11 2008) requires obtaining a permit for any waste treatment installation from the competent authority. Art. 23(4) says: "It shall be a condition of any permit covering incineration or co-incineration with energy recovery that the recovery of energy take place with a high level of energy efficiency" (further, but tentatively still, elaborated in Annex II to the Directive).

1.3 EUROPEAN JURISPRUDENCE

In the jurisprudence, the conditions of Art. 13 do not seem to have played a role with regard to the impact of the public at large or of local residents. One of the Naples waste crisis cases provides an example where it could have been important. Case C-297/08 *Commission v. Italy (2010)*, before the European Court of Justice (ECJ), concerned the proper fulfilment of obligations under Articles 4 and 5 of the 2006/12/EC Waste Directive.

These obligations are virtually identical to those of its successor, Waste Framework Directive 2008/98/EC. The Court considered that the 2006 Directive was faithfully transposed into Italian legislation. Article 178(2) of Decree no. 152/06, e.g. properly transposes Article 4 of Directive 2006/12/EC when it says:

> Waste must be recovered or disposed of without endangering human health and without using processes or methods which could harm the environment and, in particular:
> (a)...
> (b) without causing a nuisance through noise or odours.
> (c) without adversely affecting the countryside or places of special interest, protected by the legislation in force.[23]

In this particular case, the opposition of local residents was important but Italy pleaded that their opposition was a factor of *force majeure* preventing the faithful execution of its obligations under the Directive. In the next section, we will see that this argument was completely, almost off-hand, dismissed.

Although WtE incineration needs specific installations in comparison to ordinary incineration, its process of incineration is not so different or more dangerous to qualify in the category of excepted specialized installations like those for hazardous waste. Such plants often require a special location farther away from the waste source than the general principle of proximity normally would allow. Besides an exception for such specialized installations, Article 16.2 of Waste Directive 2008/98/EC only refers to "taking into account of geographical circumstances" for a designed network of disposal sites. It does not admit as an exception anything like, e.g. the wide-scale opposition of the (local) population.[24]

According to Article 16.3, the network of waste disposal installations that Member States must set up, shall enable waste to be disposed of by means of the most appropriate

23 Directive 2006/12/EC (OJ 2006 L 114/9) has no reference to (national) legislation, it only requires: "(c) without adversely affecting the countryside or places of special interest."

24 Directive 2008/ 98/EC does not differ in this respect from its predecessor, Directive 2006/12/EC.

methods and technologies "in order to ensure a high level of protection for the environment and public health."

Although there is perhaps some room for debate here, these "appropriate methods and technologies" determine the way the installations function and are operated, not to a location requirement like, e.g. an appropriate distance from residential areas.

Due to the principle of proximity, waste must be disposed of, generally speaking, as close as possible to the place where it is produced. Further location criteria determine the precise location of each individual waste disposal facility. According to the ECJ these further criteria are to be determined in the light of the objectives pursued by the applicable legislation like the 2008 Framework Directive, which include the protection of health and the environment as the core condition.[25] The Court specifies that

> ...[t]hose location criteria should accordingly relate, *inter alia*, to the distance of such sites from inhabited areas where the waste is produced; the prohibition on establishing installations in the vicinity of sensitive areas; and the existence of adequate infrastructure for the shipment of waste, such as connections to transport networks...[26]

As said above, Article 13(b) of the 2008 Waste Framework Directive (2008/98/EC) with its prohibition of causing nuisance through noise or odours may provide a reasonable opportunity to effectuate (legal) protest, but, as yet, reports do not show that the provision has served in that way. Such an explicit opportunity to protest may eventually provide some leeway to "pro-environmentalist arguments" in the sense of the Californian study referred to above.[27]

In the Naples waste crisis case C-297/08, just referred to, some attention is explicitly paid to protest by the local population against a plan to construct a waste incinerator facility "in their back garden", albeit in a rather peculiar way, and not really along the lines of the Californian argument. One of the Italian arguments advanced was that the popular sentiments against the (WtE) incinerator planned were of such a nature that the Italian authorities could not reasonably be expected to cope with them. The authorities in the Campania had used a regional law to define 18 homogeneous territorial zones for (urban) waste where they wanted to create an integrated network of waste disposal installations. The procurement contracts in question were awarded, including for the construction of seven plants producing combustible materials derived from waste (CMW) and two thermal recovery plants, to be located at Acerra and Santa Maria La Fossa,

25 In Joined Cases C-53/02 and C-217/02 *Commune de Braine-le-Château and Others* [2004] ECR I-3251, paragraph 32 & 34 and Case C-297/08 *Commission v. Italy* [2010] ECR-1, cons. 65, the reference is to the objectives of the preceding 'Framework' Directive 2006/12.

26 Case C-297/08 *Commission v. Italy, idem.*

27 *See, op. cit.*, footnote 10.

respectively. The municipalities in the region of Campania were required to have their waste treated by those companies. However, "…first, because of opposition from some of the local inhabitants concerning the sites selected and, secondly, because of the low volume of waste collected and deposited with the regional service," the plan could not really be executed.[28] Moreover, plant construction ran into delays, and flaws were detected in the design of the installations. As a result, waste accumulated to saturation point in the available landfills and storage areas because it could not be treated by the facilities in question:[29] the Naples waste crisis was an embarrassing fact. Not surprisingly, the European Commission was not impressed by the Italian efforts. The Commission held that Italy

> …is in breach of Articles 4 and 5 of Directive 2006/12, in that (i) it has not established an integrated and adequate network of disposal installations capable of ensuring self-sufficiency in waste disposal on the basis of the criterion of geographical proximity and (ii) that the situation has become a source of danger for human health and for the environment.[30]

Italy defended itself by pointing to a situation of *force majeure* blaming the non-performance of public procurement contracts by the companies involved, the opposition of the local inhabitants and criminal conduct outside its control.

None of the Italian *force majeure* arguments was accepted by the Court.[31] With regard to local opposition to the establishment of disposal installations, the Court did not give much leeway. The judgment just refers to a general settled position in ECJ jurisprudence:

> …it is settled case-law that a Member State may not plead internal situations, such as difficulties of implementation which emerge at the stage of putting a Community measure into effect, including difficulties relating to opposition on the part of certain individuals, in order to justify a failure to comply with obligations and time-limits laid down by Community law.[32]

28 *Id.*, cons. 16.

29 *Ibidem.*

30 *Id.*, cons. 29.

31 In another Naples waste crisis case, this time before the European Court of Human Rights, Italy also pleaded *force majeure* on the same grounds: *Case of Di Sarno and others v. Italy* (application no. 30765/08), Judgment of 10 January 2012, cons. 99. In rejecting the *force majeure* argument, the Strasbourg Court referred directly to the conditions of Article 23 of the Articles of the United Nations International Law Commission on State responsibility for internationally wrongful acts (cons. 111).

32 Case C-297/08 *Commission v. Italy, op. cit.*, footnote 25, cons. 85 The judgment refers in particular to Case C-45/91 *Commission v. Greece* [1992] ECR I-2509, paragraphs 20 and 21 and the more recent Case C-121/07 *Commission v. France* [2008] ECR I-9159, paragraph 72. It reminds of Article 46 of the (1969) Vienna Convention on the Law of Treaties, although that article refers to provisions of internal law only (*see also* the preceding footnote).

The Italian arguments in this particular case regarding local protest may, indeed, not be very convincing. However, the Court's total dismissal of such protest, including the arguments it used, does not particularly fit in well with steps of facilitating participation in decision-making on the siting of incinerators. It does not help to avoid nuisance for individuals living nearby the site planned or encourages to take environmental concerns or negative consequences for "human beings" in densely populated areas into consideration. The judgment anyway does not contribute much to improving the "image" of (WtE-)incinerating plants.

EU law may nevertheless provide somewhat more legal room for (local) people who do not want waste incinerator (WtE) plants next door than this judgment suggests. Article 7 of the "Charter of Fundamental Rights of the European Union," in force since 1 December 2009, says that "Everyone has the right to respect for his or her private and family life, home and communications." The provision, as yet, has not made many waves in ECJ jurisprudence and perhaps in the current state of the law even cannot have much impact.[33] However, its "sister" Article 8 of the European Convention on Human Rights has a record of at times producing remarkable, or at least promising, results for individuals who try to defend themselves against projects that have or may have a negative effect on their privacy and home (notwithstanding quite a formidable "margin of appreciation" in Article 8.2). In famous cases like the 1994 Judgment in *López Ostra v. Spain*, the European Court of Human Rights found it fit to provide protection against undue intrusions of home, private and family life. The Strasbourg Court held:

> Naturally, severe environmental pollution may affect individuals' well-being and prevent them from enjoying their homes in such a way as to affect their private and family life adversely, without, however, seriously endangering their health.[34]

The Court decided thus that even if health was not immediately threatened, the state of the environment – due to a failed waste incinerator plant in this case – could have a (very) negative impact on the quality of life of the complainants.[35] Spain was held responsible for violating the right to respect for the home and private life of the complainants under

33 Article 51 of the Charter (2000 OJ C 364/01) limits its application: the Charter articles only apply to the EU institutions and the Member States "...when they are implementing EU Law" (Art. 51.1). They do not "... establish any new power or task for the Community or the Union, or modify powers and tasks defined by the Treaties" (Art. 51.2).

34 Case of *López Ostra v. Spain* (application no. 16798/90), Judgment of 9 December 1994, Cons. 51; *see, e.g.*, Philippe Sands, 'Human Rights, Environment and the Lopez Ostra Case: Context and Consequences', European Human Rights Law Review, Issue 6, 1996, pp. 597-618; or: Richard Desgagne & Judith Hippler Bello, 'Lopez Ostra v. Spain', American Journal of International Law, Vol. 89, No. 4, 1995, pp. 788-791.

35 For a case of (aircraft) noise nuisance, *see, e.g., Hatton and others v. The United Kingdom* (application no. 36022/97), Judgment by the Grand Chamber of 8 July 2003. *See, also*, cons. 6 referring to the 2001 Chamber Judgment in this case where the Chamber, indeed, decided that there had been a violation of Art. 8.

Article 8 of the Convention. The Court found that living for years near a source of smell and smoke can impact an individual's well-being and prevent enjoying his or her home in such a way that private and family life is damaged. The prohibition of causing nuisance through noise or odours under Article 13(b) of the 2008 waste Framework Directive in the form of a similar provision of (transposed) national legislation may open an interesting opportunity for local residents (at least) seriously to plead against the issue of a permit. Perhaps also appealing to Article 7 of the EU Charter may support such a claim. Eventually an application by local residents to the Strasbourg Court may lead to a case against the operation of a (WtE) incinerator project, which is bound to cause nuisance due to (excessive) "noise and odours". But, still, this seems to be a long and demanding way with only a remote chance of success to prevent the location of such an installation by way of legal procedures. For that purpose the detailed criteria of the EIA to be fulfilled for such a project, including the right to be informed and to participate in the relevant decision-making process according to the Aarhus principles normally speaking will provide more realistic guarantees for the protection of home, private and family life than the "human rights" trajectory.

1.4 The Waste Hierarchy

Polishing up the negative image of waste incineration has not been exactly a straightforward process, but some improvement can be noticed. Notably progress in the technical sense has made it possible to realistically strengthen the requirements and conditions set in regulatory instruments, which now also provide more possibilities for public participation and opposition. Still WtE's negative reputation has continued to resonate for a long time. When a so-called "Waste hierarchy" (or "Waste pyramid") was inserted in the basic waste management legal instrument, WtE processes ended up quite at the bottom. Article 4.1 of the 2008 EC "Framework" Directive on Waste gives the following order:

> *Article 4*
> Waste hierarchy
> 1. The following waste hierarchy shall apply as a priority order in waste prevention and management legislation and policy:
> (a) prevention;
> (b) preparing for re-use;
> (c) recycling;
> (d) other recovery, e.g. energy recovery; and
> (e) disposal.[36]

36 Directive 2008/98/EC, *op. cit.*, footnote 22.

Besides assigning them this low rank in the waste hierarchy, WtE processes are further hardly explicitly referred to in the Directive. In Annex I, "Incineration on land" is mentioned as a Disposal Operation. Annex II refers under R1 to "Use principally as a fuel or other means to generate energy." A footnote to Annex II provides the technical criterion an incineration/WtE operation should fulfil in order to be classified as such a recovery operation.[37]

In 2011, the European Commission provided some (not legally binding) clarification of the R1 Energy Efficiency formula of Annex II.[38] Some more explanation of this crucial formula was asked for as it is after all to be used in order to "promote" an incinerator (or comparable plant) from a disposal operation to a recovery operation in the Waste hierarchy. The consequences of such a promotion might be considerable as is already implied by the introductory sentence in the Foreword to the 2011 Commission "Guidelines": the new 2008 Framework Directive is characterized as marking "... a shift away from thinking about waste as an unwanted burden to seeing it as a valued resource." And then, in particular, about WtE:

> With the so-called R1 Formula, it [i.e. the 2008 Directive] has introduced an incentive for municipal waste incinerators to contribute to the energy supply of industries and households. Municipal waste incinerators meeting or exceeding the energy efficiency thresholds of this formula can be qualified as facilities for the recovery of energy from waste according to the waste hierarchy.[39]

However, some serious concerns were voiced almost at the same time. In 2012, Eurostat reported that

> ...49.8% of all wastes generated in the EU were disposed of (namely, landfilled and incinerated without energy recovery) in Member States while approximately 5% were energy recovered, representing 1.3% of all energy used in the EU.[40]

---

37 The footnote states that incineration facilities dedicated to the processing of municipal solid waste only qualify "where their energy efficiency is equal to or above:
- 0.60 for installations in operation and permitted in accordance with applicable Community legislation before 1 January 2009,
- 0.065 for installations permitted after 31 December 2008."

The footnote provides then a detailed formula which "... shall be applied in accordance with the reference document on Best Available Techniques for waste incineration".

38 'Guidelines on the interpretation of the R1 energy efficiency formula for incineration facilities dedicated to the processing of municipal solid waste according to Annex II of Directive 2008/98/EC on waste', dated June 2011.

39 *Id.*, p. 2, "Foreword".

40 European Commission Indicative Roadmap of 1/2016 'Exploiting the Potential of Waste to Energy under the Energy Union Framework Strategy and the Circular Economy', p. 2.

Moreover, 14 Member States still landfilled more than 50% of their municipal waste. Hence, the Commission observed that important amounts of the energy content of non-recyclable municipal waste are being leaked from the *circular economy*. The circular economy model has become the prevailing strategic economic policy objective in the EU since the adoption of the 2 December 2015 EU Action plan for the Circular Economy.[41] In its 2016 "Indicative Roadmap" on WtE, the Commission argues still rather positively about WtE in this perspective:

> Energy and material recovering non-recyclable waste [sic] not only makes sense under the resource circularity, but also would avoid landfill emissions with a powerful climate change potential; displace the use of virgin fossil fuels and contribute to the security of energy in the EU by producing energy from a renewable source.[42]

The contribution of WtE could be strengthened still and the Commission suggests ways of making existing WtE processes more efficient. The Roadmap also notes that apparently there is a lack of clarity with respect to the waste hierarchy notwithstanding the Commission's 2011 Guidelines referred to above. A Communication to improve understanding of this concept and its implications is announced and, indeed, was issued in early 2017 under the promising title *The Role of Waste to Energy in the Circular Economy*.[43]

1.5 WtE Processes and Climate Change

Since the 1990s, in respect to the "sustainability" of waste management, still another matter required attention: incineration plants of MSW emit considerable amounts of CO_2. In its 2017 Communication on WtE, the European Commission deals with this issue more or less similarly as if it concerned an "ordinary" middle or larger industrial plant, i.e. by way of the application of as many energy efficient techniques as possible:

41 *Closing the Loop – An EU Action Plan for the Circular Economy*, COM(2015) 614 final. The Communication explains (in a footnote): "A circular economy is one in which the value of products, materials and resources is maintained for as long as possible, minimising waste and resource use." *See*, further, for a detailed account of circular economy thinking in the EU, Chapter 6 by Thomas De Römpf in this book.

42 EC Indicative Roadmap..., *op. cit.*, footnote 40, p. 1/2.

43 Communication from the Commission to the European Parliament, the Council, the European Economic and Social Committee and the Committee of the Regions, 'The role of waste-to-energy in the circular economy', Brussels, 26.1.2017 COM (2017) 34 final.

> Where waste-to energy processes are opted for, there is a need to ensure that the most efficient techniques are used: this maximizes their contribution to the EU's climate and energy objectives.[44]

There is quite a lot to gain here. In the 2017 Communication, on the basis of other studies, the European Commission reported that the amount of energy recovered from waste could rise by 29% using the same amount of waste.[45] It lists the various techniques to improve the performance of the major WtE processes and highlights the energy efficiency levels attainable by installations working in combined heat and power (CHP) mode, compared to plants merely producing either heat or electricity.[46] The Commission concludes its 2017 Communication by arguing in this respect:

> ...it is only by respecting the waste hierarchy that waste-to-energy can maximize the circular economy's contribution to decarbonisation, in line with the Energy Union Strategy and the Paris Agreement. As mentioned earlier, it is waste prevention and recycling that deliver the highest contribution in terms of energy savings and reductions in GHGs emissions.[47]

1.6 A Delicate Matter: Balancing the Availability of MSW and Incinerator Capacity

There cannot be much doubt that WtE processes in principle fit in well with the move to a (more) circular economy that is undertaken in several parts of the world, in particular in the European Union.[48] However, this generally rosy perspective for WtE is perhaps marred by the effects of the success of the switch to the circular model: the continuous availability of, in particular, MSW may get into jeopardy. Globally seen, any lack of available MSW in the near future is unlikely. Rather on the contrary, one may say, according to a 2012 World Bank study, the amount of MSW generated is expected to grow faster than urbanization rates in the coming decades, reaching 2.2 billion tons/year by 2025 and 4.2 billion tons/year by 2050.[49] However, national and regional availability will differ substantially, and here a rather serious problem lurks for the WtE sector and perhaps for waste management in general. A recent study commissioned by the European Environment Agency (EEA) shows that the incineration capacity in the EU increased between 2010 and 2015 but that it is unevenly spread in the EU: in the southern and

44 *Id.*, p. 9.
45 *Ibid.*
46 *Id.*, p. 10.
47 *Ibid.*
48 *See*, the Chapters 6 by De Römpf and 7 by Talus & Turunen in this book.
49 *What a Waste: A Global Review of Solid Waste Management*, World Bank, 2012.

eastern parts of the EU, incineration capacity is (very) limited and reliance on landfill is still widespread. At the same time, in the northern and western countries the availability of MSW is diminishing as a result of increasing recycling, reuse and prevention of waste. This "in-congruency" within the EU leads to apparently rapidly increasing shipments of waste for energy recovery.

> Total imports increased by more than a factor of four between 2008 and 2013, [...] from around 300 000 tonnes to around 1.4 million tonnes, while total exports increased six-fold in the same period, from around 380 000 to 2.3 million tonnes.[50]

Although, in view of the total amount of MSW generated in the EU this is still a relatively small amount, there is already reason for concern as we will see in the following. As a whole, the EU has no over-capacity of incineration facilities. However, some Member States clearly rely strongly (the 2017 Communication terms it "excessively") on incineration of municipal waste. This situation, the Commission argues, "...may be partly explained by high demand for heat through district heating networks, the higher efficiency of their waste-to-energy processes and high levels of social acceptance." Although this is perhaps more clearly understandable and presents the reverse of the "bad image" of incineration noted previously, the Commission notes: "Nonetheless, such high rates of incineration are inconsistent with more ambitious recycling targets."[51] This problem can be addressed, the Commission continues then to argue – in the process showing its less than great enthusiasm for WtE processes – by such rather drastic steps as introducing or increasing incineration taxes, phasing out support schemes for waste incineration (and/or supporting higher-ranking processes in the waste hierarchy), and introducing a moratorium on new facilities and de-commissioning older and less efficient ones.[52] In the EU, recovering energy from waste accounts for approximately 1.5% of the total energy consumption. This percentage should not significantly increase, the Commission expects, as more waste is directed to recycling. In fact, the amount of household and similar waste as a feedstock for WtE incineration "is expected to fall as a result of separate collection obligations and more ambitious EU recycling targets." Member States are even "advised to gradually phase-out public support for the recovery of energy from mixed waste."[53]

50 *Assessment of Waste Incineration Capacity and Waste Shipments in Europe*, Jasmina Bogdanovic *et alia*, Study commissioned by the European Environment Agency, January 2017, at p. 43. The main specific reason for this upswing in shipments seems to have been the surge of the United Kingdom as a major exporter of waste to other European countries (*id.*).

51 2017 EC Communication, *op. cit.*, footnote 43, p. 7.

52 *Id.*, p. 8. An earlier CEWEP (Confederation of European waste-to-energy plants) report, not surprisingly perhaps, shows already its total rejection of such measures, in particular tax measures, to discourage the sector (in: 'Waste-to-Energy, A Vital Link for a Circular Economy; CEWEP's Response to the Commission's Proposal on the Circular Economy', CEWEP 2014, p. 3).

53 2017 EC Communication, *op. cit.*, footnote 43, p. 6.

Presuming that, indeed, recycling increases, improving the energy efficiency of WtE processes is proposed as the most promising development. Most consideration should be given to such processes "as anaerobic digestion of biodegradable waste, where material recycling is combined with energy recovery."[54] As a consequence, the role of waste incineration as the predominant WtE option needs to be redefined. The 2017 Communication ends by warning: "When planning future investments on waste-to-energy capacity, it is essential that Member States take into consideration the risk of stranded assets," and concludes:

> The Commission remains committed to ensuring that EU funding and other public financial support is directed towards waste treatment options that are in line with the waste hierarchy and that priority is given to waste prevention, re-use, separate collection and recycling.[55]

Whether the Commission's attempt in this way to guide and direct the rather dynamic WtE market will be successful is, of course, to be seen, in particular in view of its rather delayed and (still) somewhat reluctant if not ambiguous interest in the sector, and perhaps also because of WtE's increased social acceptance.[56]

The long-awaited Communication of the Commission cannot be said to present a spectacularly profound re-evaluation of the role of WtE at the crossroads of the EU's waste and energy policy. However, it does point to an emerging new problem for the sector, following its resolution of many technical problems, and the more or less successful improvement of its rather difficult "public image".[57] As has been noted above, the incongruence between incineration capacity in the EU and the availability of solid waste to feed WtE plants leads to an already considerable increase in shipments of MSW and perhaps of non-MSW as well. This increase may prove a rather challenging development. The 2017 Communication argues that cross-border shipments of waste could help to make optimal use of capacity already available in some Member States. The Communication adds cautiously – if not hesitantly – that such exports for energy recovery "...should not necessarily be seen as contradicting the so-called principle of proximity (i.e., using the

---

54 *Id.*, p. 11.

55 *Ibid.*

56 The 2017 Communication appears some 9 years after the 2008 Framework Directive, and so does the important study *Assessment of waste incineration capacity...* (*op. cit.*, footnote 50). This study emphasizes the need for further research also in view of the limited, and often not yet sufficient, data available: "There is, in any case, a clear need for innovative and transparent assessment tools to improve coordination of incineration capacity. Non-municipal waste flows and energy recovery capacities for such wastes are heavily intertwined with municipal solid waste flows. Analysis of such flows, potentially significantly larger than municipal solid waste flows, is severely hampered by a lack of consistent data across Europe. Better data are expected to become available in the future and would merit analysis." (at p. 43).

57 Pending, of course, the results of an extensive study like the 1985 Californian study referred to earlier, *op. cit.*, footnote 10.

nearest appropriate facility) that underpins EU waste legislation."[58] Indeed, "not necessarily", but such a development may be called at least an important and perhaps risky if not outright undesirable development in the field of European environmental policy in general and European waste management policy in particular. The Commission suggests, as a perhaps rather cumbersome solution, that an LCA should be carried out to ensure "...that the overall environmental impacts, including those related to the transport of waste, do not offset the benefits."[59]

Predicting the consequences of the development of a dynamic solid waste market triggered by further WtE investments may not be simple at all for one reason because it presumes a considerable insight into the state of the relevant market, not only of MSW but also, and in particular, of industrial solid waste, and of the electricity and energy markets in general. In a 2014 report CEWEP (Confederation of European Waste-to-Energy Plants) assessing the state of the WtE market including the availability of adequate streams of MSW, warns that much more "accurate measurement, better data and transparency about what goes in and what goes out of Recycling facilities" is required. Moreover, CEWEP warns that

> Greater transparency is also necessary in order to avoid that the EU's "recycling" targets are achieved simply by shipping the waste to countries with poorer environmental and social standards than those in Europe.[60]

By 2017, in terms of measurement and transparency, the situation seems not to have improved much. The report "Assessment of Waste Incineration Capacity" commissioned by the European Environmental Agency, argues that

> Non-municipal waste flows and energy recovery capacities for such wastes are heavily intertwined with municipal solid waste flows. Analysis of such flows, potentially significantly larger than municipal solid waste flows, is severely hampered by a lack of consistent data across Europe.[61]

Upon the availability of such improved data, the place and role of WtE processes in the waste management and energy policy of the EU is probably rather urgently in need of

58 Indeed, it can be argued that the rather complicated Art. 16 of the 2008 Framework Directive that currently lays down the principle, offers enough exceptions for escape routes, but it would be exaggerated to call that provision a model of transparency (2017 EC Communication, *op. cit.*, footnote 43, p. 7).

59 2017 EC Communication, *id.*

60 The Report refers to China, in particular regarding the export of post-consumer plastic waste. This CEWEP report comments on a 2014 Commission proposal for a considerable revision of the EU waste targets (i.e., <http://eur-lex.europa.eu/legal-content/EN/TXT/?uri=CELEX:52014PC0397>).

61 *Assessment of Waste Incineration Capacity*. The authors of the study do expect that better data will become available in the future; they would merit analysis (p. 43, *op. cit.*, footnote 56).

further clarification and elaboration beyond what the 2017 Communication has provided. Notably applying the "waste hierarchy" in practice seems to prove more difficult than the ideal of the circular economy allows. Moreover, WtE is not only relevant as a waste management technique. It is also part of energy policy, including by now as a relevant (at least local) factor in respect to urban heating processes and as a means of helping to decrease the import of fossil fuels. In a city like Amsterdam, the municipal WtE plant (Afval Energie Bedrijf, AEB) in 2015 produced more than 890.000 MW electricity and contributed considerably to a rapidly growing municipal heating network.[62] In view of the dynamics of the waste-to-energy sector, it seems advisable for the Commission not to wait too long with such further clarification.

1.7 Conclusion

Generally speaking, it seems that the future for WtE is most promising. Investment figures show a "booming" sector. The increase in investment is almost worldwide, and in particular in the European Union its future seems bright. The pollution problems of the 1970s and 1980s have been addressed seriously and adequately, it seems, both in terms of the development and application of technology, and in terms of the adoption of reasonable regulation. Moreover, the well-established suspicion of the public against the incineration of MSW seems to have been overcome not in the least by, in principle, opening up accession to information and to decision-making processes regarding new plants. This includes much greater openness on siting, in particular by way of ever more sophisticated environmental impact assessment legislation up to the level of life cycle assessments of materials and services. Also, the treatment of WtE plants on equal footing with other medium- and large-sized industries and the emphasis on energy efficiency seems a good way to address the contribution of WtE processes to the problem of the accumulation of greenhouse gases in the atmosphere. The potential for energy efficiency in the sector, at least according to studies of the European Commission, seems quite promising and may further diminish its role as a greenhouse gas contributor, ánd, of course, enlarge the electricity production from MSW and waste from other sources. As we have seen, energy production from WtE processes is not uncontroversial, although there is also some reason to believe that "local" production of energy from waste scores well in the public image, a development also noted by the Commission. European Commission activity with regard to the sector has perhaps not been overwhelming, clearly leaving room or even a need for further elucidation of the place and development of WtE processes in the future, including within the circular economy strategy perspective. The uneven availability of WtE plants in the European Union and the future availability of solid waste fuel

62 2015 AEB Annual Report, at <www.aebamsterdam.nl/media/1621/aeb160623_jaarverslag-2015.pdf>. *See*, further, Chapter 4 of this book.

to feed those plants is a subject of serious concern, including of course for the sector. Moreover, the resulting and potentially increasing market of solid waste leading to growing flows of such waste in Europe, as well as from other parts of the world into Europe, is a development requiring a more adequate assessment and regulation.

2 Investment in WtE: The World Bank and Others

Stefano Silingardi

2.1 Introduction

Investments in waste to energy (WtE) are increasing throughout the world. The WtE market size was US$20.86 billion in 2015 and is expected to witness a significant growth, rising to US$37.64 billion in 2020 and to almost US$44 billion in 2024.[1]

As, at the end of the day, WtE technologies allow us to take waste and turn it into a useable form of energy, it is quite natural that these technologies are becoming a popular, very practical solution in developed markets to solve the dilemma of ongoing urbanization: a high, increasing rate of generation of all types of waste, on the one side, and an ever-increasing need for energy, on the other side. Moreover, WtE also contributes to the energy security, by making countries less dependent on fossil fuel imports[2] as well as helping them to achieve the most ambitious recycling targets (such as those set in the EU by the Waste Framework Directive).

However, many challenges still exist that hinder the development of such technologies. Social opposition and environmental concerns, for example, make WtE plants difficult to site as residents tend to reject the idea of waste processing and disposal taking place in their vicinity; on the other side, there is often concern that the development of WtE will legitimize the creation of waste and discourage greater recycling. Another challenge for WtE, especially in least developed and developing countries, is represented by poor quantity and poor quality of waste as the energy output is largely affected by the calorific energy content of the various types of waste.

Nonetheless, the most critical part of any WtE project is probably still represented by ensuring secure financing. Indeed, investments in WtE tend to be long term and capital intensive, thus making the availability of secure financing a critical concern especially in least developed and developing countries.

There are primarily five widely used and implemented technologies to create energy from a waste source, namely: incineration (e.g. the combustion of organic material such as waste with energy recovery), pyrolysis or gasification, plasma arc gasification or the

---

1 *See* at: <www.statista.com/statistics/480452/market-value-of-waste-to-energy-globally-projection/>.

2 The energy content of the waste treated by WtE plants in Europe amounts to 19% of the gas imports from Russia in 2012. *See*: Eurostat (2012), *Table nrg_124a – Imports – gas – annual data*, at: <http://ec.europa.eu/eurostat/web/energy/data/database>.

plasma gasification process, refuse derived fuel (RDF) and bio-methanation (e.g. anaerobic digestion).[3] WtE technologies based on thermal energy conversion (e.g. incineration, pyrolysis, plasma arc gasification) lead the market, accounting for 88.2% of total market revenue in 2013.[4] As it has been noted by the EU Commission in its recent *Communication on the role of WtE in the circular economy*[5] these processes encompass very different waste treatment operations, ranging from disposal and recovery to recycling, they have different environmental impacts and rank differently in the waste hierarchy. For example, processes such as anaerobic digestion which result in the production of biogas and a digestate are regarded by EU waste legislation[6] as a recycling operation. On the other hand, waste incineration with limited energy recovery is regarded as disposal.

These technologies can be applied to several types of waste, but the most common application by far is processing municipal solid waste (MSW).[7] The increasing rate of MSW processing over the forecast period is therefore anticipated to make it the most relevant fuel for the global growth of WtE. MSW increase will be mainly driven by a growing urban population, with 4.3 billion urban residents by 2025, generating about 1.42 kg/capita/day of municipal solid waste (2.2 billion tonnes per year)[8] with other factors such as industrialization, economic development, local climate and public habits such as leftover food, further contributing towards MSW generation.

Alongside the increasing rate of waste generation, other drivers for the growth of the WtE market could be summarized as high energy costs, restricted landfilling capacities and growing concerns about environmental issues. Under that last perspective, for example, both the United Nations Framework Convention on Climate Change (UNFCC)[9] and the Paris Agreement[10] requires countries to reduce their carbon output as soon as possible and to do their best to keep global warming to less than 2°C above pre-industrial levels, and make best efforts to limit warming to 1.5°C. In this context, it is unsurprising

3 World Energy Council, *Waste-to-Energy 2016*, London, 2016, pp. 12 ff. (*see*, for more detail Chapter 3, below).

4 World Energy Council, *World Energy Resources 2016*, London, 2016, p. 26.

5 European Commission, *Communication on the Role of WtE in the Circular Economy*, COM(2017) 34 final, Brussels, 26 January 2017.

6 Article 2 (6) of Commission Decision 2011/753/EU, OJ L 310 of 25 November 2011.

7 Eurostat (2013), at: <http://epp.eurostat.ec.europa.eu/portal/page/portal/waste/introduction/>.

8 World Bank, *What a Waste, A Global Review of Solid Waste Management*, Urban Development Series Knowledge Papers, March 2012, No. 15, p. 8, according to that report, there were approximately 2.9 billion urban residents who generated about 0.64 kg of MSW per person per day (0.68 billion tonnes per year) in 2002; and about 3 billion residents generating 1.2 kg per person per day (1.3 billion tonnes per year) in 2012. The increase in per capita waste generation rates is therefore significant, from 0.64 to 1.42 kg per person per day in 25 years. According to the most recent data, *see World Energy Resources 2016, supra* note 4, it is estimated that global waste generation will double by 2025 to over 6 million tonnes of waste per day. While OECD countries will reach "peak waste" before 2050, and East Asia and Pacific countries by 2075, waste will continue to grow in Sub-Saharan Africa. By 2100, global waste generation may hit 11 million tonnes per day.

9 1991 United Nations Framework Convention on Climate Change, Rio, 1771 UNTS 107.

10 2015 Paris Agreement, entered into force 4 November 2016, text available at: <http://unfccc.int/files/essential_background/convention/application/pdf/english_paris_agreement.pdf>.

that WtE projects are proving an attractive option for countries as they would help to divert waste from landfills and therefore can substantially contribute to reducing the waste volume and cut down on greenhouse gas (GHG) emissions.[11]

One of the glaring examples of this trend comes from the Middle East region, in particular among those countries participating in the Gulf Cooperation Council (GCC). Historically, these countries, which can be counted among the world's largest waste producers in terms of per capita waste generation, have placed a reliance on landfill for waste disposal. Kuwait, for example, has 18 landfills, of which 14 sites are closed and four sites are still in operation, for a total of 18 sq. km on the total land area of Kuwait, which is around 17,820 sq. km.[12] However, the recent rise in urbanization rate has not only caused an increase in the generation in all types of waste, but it has also led residential areas to the edges of landfill sites, thus causing grave danger to public health. Kuwait, therefore, has recently planned the construction of a municipal solid waste treatment facility (Kabd Municipal Solid Waste project),[13] which is expected to receive 50% of total municipal solid waste produced in the country. For the same reasons, over the past few years many of the Middle East countries have announced WtE projects. This has included Bahrain, Saudi Arabia, Oman – which is set to establish the Sultanate's first WtE project, projected to be worth approximately US$600-700 million to produce electricity for seawater desalination with a capacity of 1,000,000 tonnes per annum – and Dubai, which is planning to build a US$545 million plant that aims to be the largest in the Middle East by the second quarter of 2020.[14]

In the next sections, after a brief overview of the different traditional ways to finance WtE infrastructures and the risks that financial lenders, such as banks or equity investors, should manage, we will focus on the examination of the actual investment policies in the WtE sector of major international financial institutions and platforms, like the World Bank (WB), the World Energy Council, the United Nations Environment Programme (UNEP) and the UN Clean Development Mechanism. Finally, current trends in the Asian and European Union behaviours towards WtE will also be taken into account, for example the use of European funds to finance WtE projects.

11 By way of example, according to a recent study focused on waste management in the EU, 92 million tonnes of CO_2 emissions could be saved by 2030 in the EU-28 if all the municipal waste would be diverted from landfill. *See*: Institute for Applied Ecology (Öko-Institut) and the Institute for Energy and Environmental Research (IFEU) on behalf of the German Federal Environment Agency (UBA), *The Climate Change Mitigation Potential of the Waste Sector*, July 2015, available at: <www.umweltbundesamt.de/publikationen/the-climate-change-mitigation-potential-of-the>. For more on the relationship between waste disposal and climate change, *see also What a Waste, A Global Review of Solid Waste Management, supra* note 8, pp. 29 ff.; and World Energy Council, *World Energy Resources: Waste to Energy*, London, 2013, Ch. 7b.

12 *See* at: <www.ecomena.org/tag/solid-waste-management-in-kuwait/>.

13 *See* at: <www.ptb.gov.kw/en/Municipal-Solid-Waste-Treatment-Facility-Kabd-Location>.

14 *See* at: <www.energylawexchange.com/will-ppps-provide-catalyst-waste-energy-projects-gcc/>.

2.2 THE FINANCING OF WTE INFRASTRUCTURE PROJECTS: TRADITIONAL APPROACHES AND NEW CHALLENGES

Debt, equity and on balance sheet have been traditionally considered to be the three main options for the construction and development of WtE infrastructure projects,[15] and they still maintain their leadership. Most recently, for example, on 19 May 2016, a US$473 million loan has been secured by a private consortium for the construction of a WtE plant in Singapore, which is expected to be completed by 2019 and is designed to process 3,600 tonnes of waste per day and generate 120 MW of energy.[16]

Other instruments of financing have been explored by the recent practice. The year 2011 saw, for example, the peak of venture capital and private equity business investments in the WtE sector – including biomass – to a total of US$1 billion.[17] Since then, however, global trends of venture capital and private equity investments in WtE have been less significant: in 2012 they halved to US$500 million,[18] in 2013 they further fell to 66% or US$200 million,[19] 2014 saw a further 8% drop[20] and, finally, in 2015 these investments decreased by another 53% to US$100 million only.[21]

Indeed, investment in the WtE sector is a complex issue. Financial lenders often want to see long-term waste contracts for a project, because a primary source of income for a WtE project is the fee charged for accepting the waste – the so-called "gate fee" (or "tipping fee") – and not just the power produced.[22] This fee is due by the municipal authority on a "per tonne of waste" basis and contracts usually also require public authorities to provide a certain tonnage of waste per month and make the gate fee payable for that tonnage regardless of whether or not the authority can actually deliver it.

As has been noted, however, if it is understandable that public local authorities want to be careful as they are spending public money, it is at the same time true that this "low

15 United Kingdom, Department for Environment, Food and Rural Affairs, *Energy from Waste, a Guide to the Debate*, February 2014 (revised edition), p. 46, available at: <www.gov.uk/government/policies/reducing-and-managing-waste>.

16 *See*: <https://waste-management-world.com/a/473m-finance-for-singapores-3600-tpd-tuasone-waste-to-energy-plant>.

17 UNEP / Bloomberg New Energy Finance, *Global Trends in Renewable Energy Investment 2012*, Frankfurt School of Finance and Management, 2012, p. 63. This amazing increase (up 186% from the previous year) was largely due to two unusually big deals: Alinda Capital Partners bought a majority stake in the German biogas project developer, Agri.capital, for US$445 million dollars; and Plasco Energy, a Canadian WtE developer, raised US$143 million dollars in private equity.

18 UNEP / Bloomberg New Energy Finance, *Global Trends in Renewable Energy Investment 2013*, Frankfurt School of Finance and Management, 2013, p. 65.

19 UNEP / Bloomberg New Energy Finance, *Global Trends in Renewable Energy Investment 2014*, Frankfurt School of Finance and Management, 2014, p. 71.

20 UNEP / Bloomberg New Energy Finance, *Global Trends in Renewable Energy Investment 2015*, Frankfurt School of Finance and Management, 2015, p. 70.

21 UNEP / Bloomberg New Energy Finance, *Global Trends in Renewable Energy Investment 2016*, Frankfurt School of Finance and Management, 2016, p. 68.

22 *Energy from Waste, a Guide to the Debate, supra* note 15, p. 47.

risk approach" inevitably makes it harder for smaller companies with innovative (commercially unproven) technologies to meet the requirements of the procurement specification and steers the larger waste companies towards offering proven technology outcomes, even if these new technologies might increase the energy output of poor quality waste.[23] Moreover, massive differences still exist between developed and emerging economies in terms of commercial viability of WtE investments for public authorities. In developed markets, where waste collection is an established, taxable function of municipal authorities, public authorities are normally willing to take the risk inherent in the "gate fee"; while authorities in emerging markets may not be able to afford the fees, nor be willing or able to increase local taxes for collection. Suffice it to consider that of the estimated 122 million tonnes of waste that are used for WtE incineration – the most developed and established technology for energy recovery from waste – over 99% is treated in high-income-level countries.

All these considerations have made it particularly difficult to apply traditional commercial schemes of WtE financing in developing countries. For governments, the major implication of using established contractual models (e.g. principally through an engineering, procurement and construction contract) has been that significant public capital is required to initiate a WtE project. In that perspective, such models are a viable solution only when the government has both the necessary budget to fund the project and the necessary expertise to procure the correct partner to construct the facility. Instead, in developing countries solid waste management, except perhaps in capitals and large cities, is generally given a very low priority, with very limited funds provided. The problem is even more acute in those countries where the local taxation system is inadequately developed and users' willingness to pay for services that are irregular and ineffective is not high either.

In that perspective, recent practice for the construction and development of WtE facilities has privileged the recourse to public–private partnership (PPP) programmes. As it is well known, there is no universally acceptable definition of PPPs. Generally speaking, PPPs may refer to both informal and short-term engagements of non-governmental organizations (NGOs), the private sector, and/or government agencies that join forces for a shared objective. PPPs may also consist of more formal engagements that can either be of a short-term nature or be more complex long-term contractual arrangements between a private party and a government agency, for providing a public asset or service, in which the private party bears significant risk and management responsibility – such as those required for the construction and operation of a WtE facility.[24] Considering the specifics

---

23 *Id.*

24 World Bank Group, *Public-Private Partnerships. Reference Guide, Version 2.0,* Washington, DC, 2014, p. 14. *See, also,* the *PPP Toolkit in Municipal Solid Waste Management*, developed by the World Bank in 2000, at: <https://ppp.worldbank.org/public-private-partnership/library/private-sector-participation-municipal-solid-waste-management-guidance-pack-5-volumes>.

of the construction and development of WtE facilities, the PPP model indeed allows governments to harness the ability of the private sector to independently raise finances and gain the expertise of operators and contractors with track records in the industry. The private sector will gain an asset with secure and long-term revenue streams (waste disposal and energy output) with the payment by local public authorities of the so-called annual "gate (or tipping) fee". It is no coincidence that many Arab Gulf countries, which are, as we have seen earlier, amongst the main developers of WtE facilities nowadays, have recently enacted internal legislatives schemes aimed to regulate the PPP programmes in that field.[25] Among others, see, for example, the 2014 Kuwait PPP law,[26] the 2015 Dubai PPP Law,[27] the Saudi "Vision 2030"[28] and the anticipated PPP laws in both Oman and Qatar.

It is precisely this context, even more worsened by the 2007 economic and financial crisis, that has fostered the role of major international financial institutions and platforms, such as the World Bank, the World Energy Council, UNEP and the UN Clean Development Mechanism, as key players to support the construction and development of WtE facilities, especially in developing countries due to the traditional difficulty in attracting foreign private capital.

2.2.1 *The World Bank Group: The IFC and the Climate Investment Funds*

According to the most recent data, the World Bank Group, e.g. mainly the International Bank for Reconstruction and Development (IBRD), the International Development Agency (IDA), the International Finance Corporation (IFC), the Global Environment Facility (GEF) and the Prototype Carbon Fund (PCF), currently has some 120 active projects with solid waste management components (with an estimated 44 additional projects in the pipeline). A total of 83 projects are under supervision with US$3.9 billion in loan/grant commitments, with solid waste management being the main component in more than half of these projects.[29] Since 2000, lending by the World Bank for solid waste management projects has reached a total of US$4.5 billion, supporting 329 solid waste programmes around the globe.

---

25 T. Barbury & T. Smith, 'Will PPPs Provide the Catalyst for Waste to Energy Projects in the GCC?', 11 July 2016, at: <www.energylawexchange.com/will-ppps-provide-catalyst-waste-energy-projects-gcc/#_ftnref1>.

26 Kuwait Law No. 116 of 2014 (New PPP Law), available at: <www.ptb.gov.kw/Library/Files/Uploaded%20Files/English/PPP%20Law%20116%20of%202014%20(English).pdf>.

27 Dubai Law No. 22 of 2015, available at: <http://ogp.dubai.gov.ae/documants/pdf/ltk2odq4mzgwmw.pdf>.

28 Full text of *Saudi Arabia's Vision 2030*, 26 April 2016, at: <http://saudigazette.com.sa/saudi-arabia/full-text-saudi-arabias-vision-2030/>.

29 See at: <http://web.worldbank.org/WBSITE/EXTERNAL/TOPICS/EXTURBANDEVELOPMENT/EXTUSWM/0,,contentMDK:20239395~menuPK:497783~pagePK:210058~piPK:210062~theSitePK:463841,00.html>.

Generally, these projects combine the financing of infrastructure and advisory services, as well as offer training in sophisticated, behaviour-changing reuse and recycling programmes.[30]

Being the only multilateral development bank with a global reach that engages directly with the private sector,[31] it is, in particular, the IFC which, according to its specific mandate, has developed major experiences in financing projects in the WtE sector as well as a specific competence in providing technical assistance and advice to governments and businesses for the implementation of PPPs transactions. More specifically, over the period 2005-2015 the IFC has invested more than US$13 billion in the area of climate-smart projects, to which WtE facilities partially contribute in achieving a net reduction of GHG emissions, totalling 103 investments in 31 countries in the fiscal year 2015 alone.[32] Within that area, over the last 10 years, US$135 million has been committed to transactions in the solid waste management sector (through equity, quasi-equity and senior or convertible loans), with a specific focus on middle-income countries, such as Turkey, Brazil, China and Mexico.

One of the major examples of the IFC involvement in this area through a PPP model is represented by its participation in an integrated waste management project for four islands in the capital region of the Maldives Islands, which started in 2011 and whose cost was estimated at US$50 million.[33] The PPP structure, awarding a 20-year build–operate–transfer (BOT) project,[34] consisted in a tripartite agreement between the bidder (the Tatva Global Renewable Energy Ltd.), the Malé City Council and the Maldives Ministry

30 *See* at: <www.worldbank.org/en/news/feature/2016/03/03/waste-not-want-not—solid-waste-at-the-heart-of-sustainable-development>.

31 M.J. Langer, 'Key Instruments of Private Environmental Finance Funds, Project Finance and Market Mechanisms', in P-M. Dupuy & J.E. Viñuales (Eds.), *Harnessing Foreign Investment to Promote Environmental Protection. Incentives and Safeguards*, Cambridge, Cambridge University Press, 2013, p. 156.

32 The three main drivers of WtE facilities are considered to be: to avoid methane emissions by diverting waste from landfills, displacing fossil fuel power generation and the contribution to recycled materials. For more on the relationship between waste-to-energy and the IFC climate smart initiative, *see* at: <www.ifc.org/wps/wcm/connect/industry_ext_content/ifc_external_corporate_site/industries/financial+markets/trade+and+supply+chain/gtfp/gfm-tsc-gtfp-ee-info>.

33 Others IFC transactions in the WtE sector include a US$2.5 million dollars quasi-equity in June 2013 in the Karachi Organic Energy Limited (KOEL), a large-scale WtE project, based on anaerobic digestion technology, in the Landhi Cattle Colony, Karachi, whose total cost was US$70 million; an investment of US$20 million of warrants in January 2012 at the Hold Co level of the C&G Environmental Protection Company, a leading WtE investor in and operator of 10 WtE projects in China, with the aim to support the company's expansion plans by exploring new greenfield projects; and a US$2.2 million equity investment in the development and management of a 25-year WtE concession in the Western Province of Sri Lanka by the RenewGen Enviro Ventures India, a company in which the IFC had invested as well a US$2.8 million quasi-equity at Hold Co level.

34 Concessions, Build–Operate–Transfer (BOT) projects, and Design–Build–Operate (DBO) projects are types of PPPs that are output focused. BOT and DBO projects typically involve significant design and construction as well as long-term operations, for new projects (greenfield) or projects involving significant refurbishment and extension (brownfield). In particular, a BOT project is typically used to develop a discrete asset rather than a whole network and is generally entirely new or greenfield in nature (although refurbishment may be involved). In a BOT project, the project company or operator generally obtains its revenues through

of Finance as guarantor.[35] According to its terms, the Agreement provides that private operator's revenue depends on the user charges from construction waste, sales of recyclables, sale of treatment by-product (energy/compost/gas) and the "tipping fee". The Government of Maldives, which has guaranteed a waste volume of 200 tonnes per day, has set the tariffs of the service and therefore carries the tariff collection risk. Moreover, the Agreement provides the obligation to comply with the Unites States Environmental Protection Agency 1998 Guidelines for Ecological Risk Assessment[36] as well as with the World Bank Performance Standards for Private Sector Activities.[37]

Amongst others, the most recent IFC investments in WtE include: (a) an investment, approved on 7 March 2016, comprising a US$30-35 million A Loan for IFC's own account, up to US$20 million of equity/quasi-equity investment by IFC, up to US$100 million syndicated B Loans and Parallel Loans to support the expansion of the Heksagon Kati Atik A.S., the first private solid waste management company that converts MSW into commercial organo-mineral fertilizer in Turkey;[38] (b) a senior loan, of US$16 million, approved on 20 May 2015, to Foxx URE BA Ambiental Ltda for the construction of a greenfield WtE facility in the municipality of Barueri, São Paulo;[39] and, finally, (c) an investment of up to US$60 million Convertible Loan, approved on 14 January 2016, to support the WtE business growth in China of the Canvest Environmental Protection Group Company Limited, a pure-play WtE provider focused on the development, management and operation of WtE plants, which was listed on the Hong Kong stock exchange in 2014.[40]

The normative framework of the IFC is supplemented by the environmental and social review of each project, which is expected to be designed and operated in accordance with applicable IFC Performance Standards (PS), the World Bank Group's Envir-

---

a fee charged to the utility/ government rather than tariffs charged to consumers. *See* at: <https://ppp.worldbank.org/public-private-partnership/agreements/concessions-bots-dbos>.

35 *See* at: <www.globalislands.net/userfiles/Maldives7.pdf>.

36 *See* at: <www.epa.gov/risk/guidelines-ecological-risk-assessment>.

37 *See* at: <http://siteresources.worldbank.org/EXTSAFEPOL/Resources/OPBP4.03GNApr22Webclean.pdf?resourceurlname=OPBP4.03GNApr22Webclean.pdf>.

38 *See* at: <http://ifcextapps.ifc.org/ifcext/spiwebsite1.nsf/78e3b305216fcdba85257a8b0075079d/4639ef1bd90-b25aa85257d8d006a6fd3?opendocument>.

39 *See* at: <http://ifcextapps.ifc.org/ifcext/spiwebsite1.nsf/78e3b305216fcdba85257a8b0075079d/4a4bc91ab215684085257dfc0077c3c2?opendocument>.

40 *See* at: <http://ifcextapps.ifc.org/ifcext/spiwebsite1.nsf/78e3b305216fcdba85257a8b0075079d/ad3d3f20f42a072885257f40005fc630?opendocument>.

onmental, Health and Safety (EHS) General Guidelines[41] and the EHS Guidelines for Waste Management Facilities.[42]

The IFC Performance Standards are directed towards clients, providing guidance on how to identify risks and impacts, and are designed to help avoid, mitigate, and manage risks and impacts as a way of doing business in a sustainable way, including stakeholder engagement and disclosure obligations of the client in relation to project-level activities. In the case of its direct investments (including project and corporate finance provided through financial intermediaries), IFC requires its clients to apply the PS to manage environmental and social risks and impacts so that development opportunities are enhanced. Each project, according to its specific features, shall comply with one or more of the eight PSs listed:[43] Assessment and Management of Environmental and Social Risks and Impacts; Labour and Working Conditions; Resource Efficiency and Pollution Prevention; Community Health, Safety, and Security; Land Acquisition and Involuntary Resettlement; Biodiversity Conservation and Sustainable Management of Living Natural Resources; Indigenous Peoples Cultural Heritage.

Instead, the General EHS Guidelines and the EHS Guidelines for Waste Management Facilities are technical reference documents with general and industry-specific examples of Good international industry Practices that members of the WB Group are requested to apply once involved in a Project. In particular, they should be tailored to the hazards and risks established for each project on the basis of the results of an environmental assessment in which site-specific variables, such as host country context, assimilative capacity of the environment, and other project factors, are taken into account; and contain the performance levels and measures that are generally considered to be achievable in new facilities by existing technology at reasonable costs. By way of example, they contain specific guidance on air emissions, ash and other residuals with regard to incineration facilities. They also provide that, to maximize the usability of end products, waste should not be accepted that contains organics that are contaminated by potentially hazardous chemicals (e.g. chlordane and other pesticides, heavy metals and metalloids) and/or pathogenic substances and micro-organisms that will not be rendered harmless by the process or may constitute a health or environmental risk.

---

41 World Bank Group, *Environmental, Health and Safety ("EHS") General Guidelines*, 30 April 2007, *see*, in particular, Section 1.6 on "Waste Management", available at: <www.ifc.org/wps/wcm/connect/554e8d80488658e4b76af76a6515bb18/Final%2B-%2BGeneral%2BEHS%2BGuidelines.pdf?MOD=AJPERES>.

42 World Bank Group, *Environmental, Health, and Safety Guidelines for Waste Management Facilities*, 10 December 2007, available at: <www.ifc.org/wps/wcm/connect/1cd72a00488557cfbdf4ff6a6515bb18/Final+-+Waste+Management+Facilities.pdf?MOD=AJPERES>.

43 IFC, *Performance Standards on Environmental and Social Sustainability*, 1 January 2012, available at: <www.ifc.org/wps/wcm/connect/115482804a0255db96fbffd1a5d13d27/PS_English_2012_Full-Document.pdf?M OD =AJPERES>.

In 2008, the World Bank jointly with the four Regional Development Banks[44] established the Climate Investment Funds (CIFs) to promote international cooperation on climate change and support progress towards the future of the climate change regime. The CIFs consist of the US$5.6 billion Clean Technology Fund (CTF) and the US$2.7 billion Strategic Climate Fund (SCF).

The CTF is intended to promote scaled-up financing for the transfer of low-carbon technologies with a significant potential for long-term GHG emissions savings.[45] The World Bank is the Trustee and Administrating Unit of the CTF, while the Regional Development Banks are the implementing agencies for CTF investments. The CTF also includes a "sunset clause", which stipulates that necessary steps to conclude CTF operations shall be taken once a new UNFCCC financial architecture takes effect. Should UNFCCC negotiations result in a renewed mandate for the CTF, operations may continue with appropriate adjustments. In the WtE sector, the CTF has been recently used, with a decision approved on 27 June 2011, to co-finance for an amount of up to US$22.46 million the establishment by the European Bank for Reconstruction and Development (EBRD) of a US$100 million Kazakhstan Waste Management Facility, which is intended to provide debt financing to municipal or private operators developing a pipeline of municipal WtE projects in Kazakhstan that is expected to result in 200,000 tonnes of avoided CO_2 per annum at the end of the four-year period of 2015-2018 with a lifetime of 20 years, and up to 10 MW of installed capacity.[46]

The SCF serves as an overarching framework to support three targeted programmes with dedicated funding to pilot new development approaches or to scale-up activities aimed at a specific climate change challenge through targeted programmes. These are: (a) the Forest Investment Program (FIP), approved in May 2009, which aims to support developing countries' efforts to reduce emissions from deforestation and forest degradation by providing scaled-up financing for readiness reforms and public and private investments; (b) the Pilot Program for Climate Resilience (PPCR), approved in November 2008, with the objective to pilot and demonstrate ways to integrate climate risk and resilience into core development planning, while complementing other ongoing activities; and (c) the Program for Scaling-Up Renewable Energy in Low Income Countries (SREP), approved in May 2009, which seeks to create new economic opportunities and increase energy access through the production and use of renewable energy. In the WtE sector, in 2014, a US$7.9 million SREP allocation has been approved for the Extended Biogas Pro-

44 The African Development Bank (AfDB), the Asian Development Bank (ADB), the European Bank for Reconstruction and Development (EBRD) and the Inter-American Development Bank (IDB).

45 Currently, over US$3.8 billion (66% of CTF resources) is approved and under implementation in clean technologies such as renewable energy, energy efficiency, and clean transport. This is expected to leverage another US$38 billion in co-financing.

46 *See*: <www-cif.climateinvestmentfunds.org/projects/waste-management-framework-kwmf>.

gram to promote large off-grid biogas energy generation in Nepal.[47] The programme, which is co-financed by the International Bank for Reconstruction and Development (IBRD) for an amount up to US$35.5 million is expected to leverage US$126.4 million in co-financing and will address high initial cost and credit financing barriers for the use of organic waste for renewable energy production, providing financing and technical assistance services for the planning and implementation of 160,000 biogas plants.

2.2.2 *WtE Investments through the UN Clean Development Mechanism*

The Clean Development Mechanism (CDM) is one of the three "flexible mechanisms" introduced by the 1997 Kyoto Protocol[48] in order to help parties included in Annex I (e.g. developed countries, currently 42 States plus the EU) in achieving compliance with their quantified emission limitation and GHG reduction commitments. The first mechanism, the International Emission Trading (IET) permits these countries to transfer parts of their "allowed emissions" ("assigned amount units"). The second, the Joint Implementation (JI), allows Annex I countries to transfer among themselves carbon credits generated by low-carbon projects undertaken within other Annex I countries. Finally, the CDM, which is sketched in Article 12 of the Kyoto Protocol, allows Annex I countries to meet part of their emission reduction commitments by buying Certified Emission Reductions (CERs) units from CDM emission reduction projects in developing countries. All three mechanisms are used for WtE investments as will be indicated below. The CDM is the most important one of the three.

Under the CDM mechanism, Annex I parties are allowed to implement projects that reduce emissions in non-Annex I parties (e.g. developing countries). In return, Annex I parties receive CERs when the project activity assists the host country in their development process and climate change mitigation. As GHG abatement costs in developing countries are expected to be much lower than in industrialized countries, the CDM gives the Annex I country the opportunity to achieve their abatement targets at lower costs while providing sustainable development projects with a mitigation potential to the host country.[49] Companies can also purchase CERs to contribute towards their own emission

---

47 *See*: <www-cif.climateinvestmentfunds.org/sites/default/files/PAD_SREP%20Nepal%20Extended%20Biogas%20project_board_approved_0.pdf>.

48 1997 Kyoto Protocol to the United Nations Framework Convention on Climate Change, 2303 UNTS 148. Currently, it has 192 Parties (191 States and the European Union).

49 *Id.*, art. 12. *See*, in particular, par. 3: "Under the clean development mechanism: (a) Parties not included in Annex I will benefit from project activities resulting in certified emission reductions; and (b) Parties included in Annex I may use the certified emission reductions accruing from such project activities to contribute to compliance with part of their quantified emission limitation and reduction commitments under Article 3, as determined by the Conference of the Parties serving as the meeting of the Parties to this Protocol." For more on the CDM, *see* D.M. Firger, 'The Potential of International Climate Change Law to Mobilise Low-Carbon Foreign Direct Investment', in *Harnessing Foreign Investment to Promote Environmental Protection. Incentives and Safeguards*, *supra* note 31, pp. 189-194.

reduction targets under mandatory emissions trading schemes, such as the EU Emissions Trading Scheme (ETS) or voluntary schemes.

In particular, private investors may be involved through primary or secondary CDM markets. In broad terms, investments in the primary market mean that the buyer receives CERs directly from the project developer (and in this case the investor would require a Letter of Approval from an Annex I country). Investment in the secondary market means that the buyer receives CERs from a third party that is not the project entity, but a secondary market entity. The main forms of investment possibilities in the CDM market are the following. First, equity investment into the project company that operates (in part or full) the CDM project, in that case the investor is enabled to share in other profit streams generated by the project, such as electricity sales. Second, investment into a carbon fund that will return either CERs or a financial return from the profits generated from the sale of CERs. Third, purchase of CERs directly from the project company, but without equitable investment in the company; this is referred to as off-taking the CERs in the primary market. And, finally, purchase of CERs from a third party entity (secondary market purchase).

Generally, in addition to being a party to the Kyoto Protocol, there are two basic requirements that countries must meet in order to participate in the CDM: (a) voluntary participation, (b) the establishment of a National CDM Authority (which have, among its other tasks, the capacity to approve private entities' participation in CDM projects). Moreover, industrialized countries must meet several further stipulations, such as the establishment of the assigned amount under Article 3 of the Protocol, a national system for the estimation of GHG, a national registry, an annual inventory, and an accounting system for the sale and purchase of emission reductions.

Any proposed CDM project has to use approved baseline and monitoring methodologies. The baseline for a CDM project activity is

> ...the scenario that reasonably represents the anthropogenic emissions by sources of GHG that would occur in the absence of the proposed project activity. A baseline shall cover emissions from all gases, sectors and source categories listed in Annex A within the project boundary.[50]

In other words, a baseline for a CDM project activity is a hypothetical reference case, representing the volume of GHG that would have been emitted if the project were not implemented.[51] Therefore, the baseline can be used to determine: (a) whether a CDM project activity respects the "additionality" requirement (e.g. that the GHG emissions after implementation of a CDM project activity are lower than those that would have

50 3/CMP.1, Ann., par. 44, available at: <http://cdm.unfccc.int/EB/rules/modproced.html>.

51 *See*: <www.cdmrulebook.org/83.html>.

occurred in the most plausible alternative scenario to the implementation of the CDM project activity);[52] and (b) the volume of additional GHG emission reductions achieved by a project activity. Instead, a monitoring methodology is the means to calculate the actual emission reductions from the project, taking into account any emissions from sources within the project boundary. A monitoring methodology sets out how project proponents should develop and implement a monitoring plan for a particular project type, in order to gather the data required to calculate emission reductions from the project.[53]

A broad range of projects are eligible for CDM accreditation, with the notable exceptions of nuclear power and avoided deforestation projects. Among them, the construction and development of WtE facilities has traditionally represented one of the most relevant activities in the area of CDM waste management projects, with a significant number of projects registered and approved in the last years. These include both biomass and WtE applications – which range from generation of electricity from bagasse and biomass, composting of solid waste to replacing conventional fuel with biomass – or capture landfill gas, which contains a large amount of methane (a strong GHG).

In the Municipal Solid Waste to Energy Project by Western Power Company (Pvt) Ltd., for example, the project activity involved the setting up of an MSW power generation project in the Kolonnawa area at Meethotamulla, Sri Lanka, designed to process around 700 tonnes per day of MSW through the process of gasification, which prior to the project was being dumped in open landfill deposits without capture of landfill gas, resulting in methane emissions due to anaerobic decomposition of MSW. In particular, a 12 MW steam turbine generator unit will be installed for generating electricity from the gasification of waste, and the electricity generated will be delivered to the national grid. The baseline scenario provides for a reduction of GHG emissions in the following ways: (i) by the displacement of the electricity generated from fossil fuel-fired plants connected to the national grid in the baseline scenario; and (ii) by avoidance of methane generation in the open landfill by utilizing the MSW waste for power generation. Also indicated in the project design document is that the estimate of annual average and total GHG emission reductions of the project activity over the fixed crediting period of 10 years are expected to be 139,829 and 1,398,295 tonnes CO_2, respectively. According to Article 12, par. 2 of the Kyoto Protocol, the project activity is expected to contribute to the sustainable development of Sri Lanka (the hot country) in terms of economic, social, technical and environmental criteria.[54]

52 3/CMP.1, Ann., par. 43.

53 *CDM Methodology Booklet*, November 2016, at: <http://cdm.unfccc.int/methodologies/documentation/1611/CDM-Methodology-Booklet_fullversion.pdf>.

54 CDM Executive Board, *Project Design Document, Municipal Solid Waste to Energy Project by Western Power Company (Pvt) Ltd*, at: <http://cdm.unfccc.int/filestorage/a/t/PL5VO9DE1Q8TMB3FRYI7KA6WJ4H0 CS.pdf/PDD_Western%20Power.pdf?t=eVJ8b2l3bzlwfDDGQ0QGKEzSqTIdmAlVq1Q->.

2.2.3 The World Energy Council and UNEP

Although not specifically involved in direct financing activities, both the World Energy Council and UNEP perform relevant activities in the WtE's investment area.

The "Energy, Climate, and Technology Branch" leads one of the key focus areas of UNEP's climate change sub-programme: the mitigation through reduction of energy-related GHG emissions and the rapid adoption of low-carbon technologies. It does so in particular in collaboration with the Frankfurt School of Finance & Management, and by way of different types of investment sources: debt, equity, asset finance, public market investments, venture capital and private equity.[55] Formed in 1923, the World Energy Council is now the UN-accredited global energy body, representing the entire energy spectrum, with more than 3,000 member organizations located in over 90 countries and drawn from governments, private and state corporations, academia, NGOs and energy-related stakeholders. Besides a number of partnerships that it has forged over the years with a number of corporations, operators of energy plants, private associations etc., today it is more active in the energy (including the WtE) sector.[56] The World Energy Council each year publishes the *World Energy Resources' Report* with details, analysis and data set, covering more than 180 countries, aimed to support governments, the private sector and academia to better understand the reality of the energy sector and the resource developments. With regard to the WtE sector, in addition to analysis and information concerning the environmental impacts, socio-economic benefits and costs inherent to the different WtE technologies available, the report also provides useful information about current trends in the WtE market, investments, costs connected to the construction and implementation of WtE facilities, and issues linked to the influence of WtE on energy security. More specifically, even if it considers that WtE still remains a costly option for waste disposal and energy generation in comparison with other established power generation sources, landfilling – which is the most financially effective way of waste disposal – could be coupled with WtE; the *2016 World Energy Resources' Report* enlightens us with the information on the existence of several different drivers that will have a positive impact on WtE market development. These include growing use of renewable energy resources, increasing amounts of waste generation globally, waste management regulations, taxes and subsidies, climate change policies to curb GHG emissions, technological advancements, new global trends such as low fossil fuel prices, environmental degradation, circular economy, green business models, industrial symbiosis (companies that work in partnerships to share resources); and improved public perception of WtE.[57]

---

55 The Frankfurt School of Finance & Management with UNEP Collaborating Centre for Climate & Sustainable Energy Finance (the Centre) each year, since 2007, publishes a report in *Global Trends in Renewable Energy Investments*, with data and statistics for the different types of renewable energy – from solar and wind, to small hydro, biofuels, geothermal, marine and biomass and WtE.

56 *See* at: <www.worldenergy.org/wec-network/global-partners/>.

57 *World Energy Resources 2016, supra* note 4, p. 28.

2.3 Regional Approaches to Investments in WtE: Asia and Europe

From a regional perspective, Europe, led by UK, Germany, Italy and France, is the largest market for WtE technologies, with about 88 million tonnes of waste treated in ca. 480 WtE plants, which account for 47.6% of the total market revenue in 2013.[58] Cumulative investment in municipal waste management over the 10 years of 2011-2020 is estimated at US$87 billion.[59] Other European countries, such as Denmark and Sweden, are among the countries that, worldwide, have most successfully implemented WtE policies. For example, currently only 1% of the Sweden's waste ends up in landfills, half of it is recycled and 49% is burned in WtE facilities, up from 39% in 1999.[60] At the same time, Europe is also the most sophisticated WtE market, with technology to utilize anaerobic-based biological processes gradually improving and possibly replacing composting as the most important step for biodegradable waste treatment generated from MSW.[61] According to a European Commission study, in 2014 approximately 1.5% of the EU's total final energy consumption was met by recovering energy from waste through incineration, co-incineration in cement kilns and anaerobic digestion (e.g. around 676 petajoules for a year).[62] However, the Commission study estimates that if proven techniques and supporting measures are properly implemented, the amount of energy recovered from waste could rise by 29% to 872 PJ/year, using exactly the same amount of waste as feedstock.

The Asia-Pacifc region, with Japan, Korea and China accounting for nearly 90% of the market, followed by Malaysia, Singapore, Thailand and Indonesia, will instead register the fastest growth (at a compound annual growth rate of 7.5%), driven by government initiatives in China and India and higher technology penetration in Japan. Even if the market is still dominated by Japan, which uses up to 60% of its solid waste for incineration and accounts for approximately 30% of the global WtE market, the fastest market growth has been witnessed in China, which has more than doubled its WtE capacity in the period 2008-2015.[63] More exactly, 90% of the total estimated US$20 billion investment in the WTE industry in Asia over the period 2011-2020 will arise from East Asian countries – specifically from China – where it is supported by national governments as a

---

58 *Id.*, p. 27.

59 Colsweep, *Report: Global Analysis of the Waste-to-Energy Field*, 2013, p. 23. *See also* CEWEP, *Waste-to-Energy Plants in Europe in 2014*, March 2016, available at: <www.cewep.eu/information/data/studies/m_1488>.

60 D. Haugen, 'Is Burning Garbage Green? In Sweden, There's Little Debate', *Midwest Energy News*, 17 October 2013, at: <www.midwestenergynews.com/2013/10/17/is-burning-garbage-green-in-sweden-theres-little-debate/>.

61 The key driver is the Council and EP Directive 2008/98/EC, OJ 2008 L 312. Another important driver for the WtE plants market in Europe is represented by the Council Directive 99/31/EC of 26 April 1999, OJ 1999 L 182, as it has resulted in the diversion of waste from landfills to WtE plants.

62 *Communication on the Role of WtE in the Circular Economy, supra* note 5, p. 8.

63 *Global Analysis of the Waste-to-Energy Field, supra* note 59, p. 20.

technical and economical option to dispose waste.[64] Chinese WtE plants that have either been operational or under construction are expected to number over 300 by the end of 2016 up from only 85 in 2008, mostly driven by favourable environmental regulations, with aggregate processing volume exceeding 35 million tonnes, making China the country with the world's largest amount of waste incineration.[65] Under this perspective, it is noteworthy that in February 2016 the Chinese government announced a project for the construction, in the mountainous region on the outskirts of Shenzhen, of the world's largest WtE plant. The plant is expected to incinerate 5,000 tonnes of rubbish per day, approximately one-third of the waste generated by the city's 20 million inhabitants each year.[66]

In this context, it is therefore not unexpected that the Asian Development Bank (ADB) and, on the other side, the EBRD and the European Investment Bank (EIB), have undertaken the most interesting initiatives in financial assistance to the construction and development of WtE facilities at the regional level.[67]

The ADB was conceived in the early 1960s as a financial institution. The bank focuses on delivering projects in developing member countries that create economic and development impact. It was established on 19 December 1966 and admits both the members of the United Nations Economic and Social Commission for Asia and the Pacific (UNESCAP, formerly the Economic Commission for Asia and the Far East or ECAFE) and non-regional developed countries. From 31 members at its establishment, ADB now has 67 members, of which 48 are from within Asia and the Pacific and 19 are from other parts of the globe. At the end of 2015, Japan holds the largest proportion of shares at 15.6%. The United States holds 15.5%, China holds 6.5% (being the first borrowing shareholder), India holds 6.3%, Australia 5.8% and Indonesia 5.5%.[68]

Beyond traditional assistance delivered through direct loans[69] and/or technical assistance, in April 2007, ADB established the Clean Energy Financing Partnership Facility

64 *Id.*

65 *See* at: <www.renewableenergyworld.com/articles/2015/04/chinese-waste-to-energy-market-experiences-rapid-growth-during-last-five-years.html>. For a more in-depth analysis of the WtE technologies in China, *see*: Suyun Xu, Hongfu He & Liwen Luo, 'Status and Prospects of Municipal Solid Waste to Energy Technologies in China', *in* O.P. Karthikeyan, K. Heimann & S.S. Muthu (Eds.), *Recycling of Solid Waste for Biofuels and Bio-Chemicals*, Singapore, Springer, 2016, pp. 31-54.

66 *See* at: <www.dezeen.com/2016/02/04/worlds-largest-waste-to-energy-plant-shenzhen-china-schmidt-hammer-lassen-gottlieb-paludan/>.

67 Albeit not comparable with the ADB and the EBRD both in terms of economic involvement and operational scope, there are other regional initiatives in the WtE sector. Among others, *see* for example the Caribbean Centre for Renewable Energy and Energy Efficiency, which implements, in cooperation with financial institutions and private investors, activities directed to mitigate financial barriers for investments in small-, medium- and large-scale renewable energy and energy efficiency projects and businesses (including the WtE sector) in the Caribbean Community (CARICOM) region.

68 <www.adb.org/sites/default/files/institutional-document/218696/oi-appendix1.pdf>.

69 *See*, for example, the ADB's decision, adopted on 10 December 2012, to lend USUS$200 million to the Dynagreen Environmental Protection Group (which is the environmental and infrastructure branch of the

(CEFPF)[70] to help provide financing to developing member countries to improve energy access and security and transition to low-carbon economies through cost-effective investments in technologies and practices that result in GHG mitigation. Among others, activities eligible to receive CEFPF resources include biomass/biofuel/biogas, carbon capture and storage, energy-efficient buildings and end-use facilities, clean energy power generation, transmission and distribution, and WtE projects (grate incineration competitive/ commercial technology, and advanced flue gas emission control). CEFPF resources are also intended to finance policy, regulatory, and institutional reforms that encourage clean energy development.

More specifically, about 30% of CEFPF's resources will be used for standalone technical assistance projects and direct charges that fund consulting services and related equipment and works needed to achieve technical assistance and direct charges objectives. About 70% will be used for concessional financing and grant components of investments and may also be used to procure equipment and works based on advanced technologies, back-financing mechanisms or risk-sharing facilities to promote clean energy, carbon capture and storage and services to lower barriers, such as biomass, biofuel and biogas technologies; WtE projects; or energy-efficient district heating, transport and street-lighting.[71]

At EU level, an initial point should be enlightened: that is, the distinction between financial support towards more sustainable waste management systems in general, on the one hand, and, on the other hand, specific financial support to the construction and development of WtE facilities. The first category of investments are mainly developed through the co-financing of the Cohesion Policy funds – in particular, the European Regional Development Fund and the Cohesion Fund – with specific preconditions that should be met to ensure that new investments in the waste sector are in line with waste management plans designed by Member States to meet their preparation for reuse and

---

Beijing State-Owned Assets Management Co.) to build at least nine WtE plants in small- and medium-sized cities of China to turn solid waste into electricity, with the objective to generate about 610 GW of electricity annually by 2018. *See* documents at: <www.adb.org/projects/46930-014/main#project-documents>.

70 CEFPF is composed of the Clean Energy Fund (CEF), the Asian Clean Energy Fund (ACEF), the Carbon Capture and Storage Fund (CCSF) and the Canadian Climate Fund for the Private Sector in Asia (CFPS). In 2015, CEFPF's total project allocations of $9.4 million are expected to leverage about US$197 million of ADB clean energy investments, supporting ADB's institutional annual target of US$2 billion. Included are about US$5 million and US$132 million of private sector and non-private sector clean energy investments, respectively. The total cumulative CEFPF project allocation is currently at US$138 million, leveraging US $2.16 billion in ADB clean energy investments and resulting in a leverage ratio of 1:28. Clean energy funds aim to leverage US$4 billion clean energy investments in developing member countries by 2020, US$1.2 billion for private sector investments and US$1.2 billion for non-private sector investments. *See*: Asian Development Bank, *Clean Energy Financing Partnership Facility, Annual Report 2015*, pp. 3 ff., available at: <www.adb.org/sites/default/files/institutional-document/182927/cefpf-annual-report-2015.pdf>.

71 *Id.*, App. 1, p. 29.

recycling targets. According to the circular economy action plan, this means that investments in treatment facilities for residual waste, such as extra incineration capacity, would only be granted in limited and well-justified cases, where there is no risk of overcapacity and the objectives of the waste hierarchy are fully respected.[72] Investments in that specific area may also be channelled through other EU financing mechanisms, such as the European Fund for Strategic Investment (EFSI), that have an important role to play in attracting private financing to the best and most "circular" solutions for waste management through loans, guarantees, equity and other risk-bearing mechanisms.

With regard to the specific area of WtE, there are at first available EU funds (e.g. Horizon 2020,[73] but also Cohesion Policy funds) conceived to support research and innovation and bring advanced energy-efficient technologies to the market. However, the most important role in support of the construction and development of WtE facilities is played by the EIB and the EBRD; the specific features of these two institutions clearly influence their approach. More precisely, the EIB, established in 1958 under the Treaty of Rome, as the European Union's international finance institution to support EU policy, whose shareholders are the EU Member States, mainly conceives investments in the WtE area as a tool to allow EU Member States to comply with European environmental standards on waste disposal and energy policy. The EBRD, founded in 1991, operates under the specific light of a multilateral developmental investment bank, therefore using investments as a tool to build market economies beyond the European Continent.[74] Moreover, despite its public sector shareholder, the latter invests mainly in private enterprises together with commercial partners.

In particular, while the EBRD has mainly acted in respect to WtE through the granting of direct loans to commercial partners (such as banks) in developing member countries,[75] the EIB has chosen a more complex approach. Over the last 5 years, the EIB has

---

72 *Communication on the Role of WtE in the Circular Economy, supra* note 5, p. 5.

73 *See*, in particular, the "InnovFin – EU Finance for Innovator", which is a joint initiative launched by the EIB in cooperation with the European Commission under Horizon 2020, available at: <www.eib.org/products/blending/innovfin/>.

74 Initially focused on the countries of the former Eastern Bloc, it further expanded and is now active in more than 30 countries from central Europe to central Asia and the southern and eastern Mediterranean (the so-called SEMED region). Egypt and Morocco originally joined the EBRD as shareholders in 1991. Jordan and Tunisia became members at the end of 2011. The EBRD has been investing in these four countries since September 2012. The EBRD is owned by 65 countries from all five continents plus the EU and the EIB, with the biggest shareholder being the United States. *See* at: www.ebrd.com/shareholders-and-board-of-governors.html.

75 *See*, for example, the recent decision of 15 May 2015 to provide a US$13 million loan to Greater Amman Municipality to help to implement a comprehensive landfill-gas (LFG) recovery system, the first of its kind in Jordan, designed and constructed with gas-collection technology, which will be used to generate electricity for delivery to the national grid, replacing electricity produced by grid-connected power plants that previously used heavy fuel oil. *See*: <www.ebrd.com/news/2015/ebrd-helps-jordan-transform-waste-into-energy-.html>. *See also* the decision of 26 October 2016 to provide a US$55 million loan to Turkish lender Isbank to finance private investments in renewable energy and resource efficiency projects, including WtE

indeed provided almost EUR 1.2 billion for investment to improve treatment and disposal of waste in 13 countries across Europe. It has done so through direct loans – such as with the EUR142 million loan agreed, on 3 March 2016, to support the construction of three WtE facilities in Cardiff, South Wales, which are expected to handle 350,000 tonnes of waste annually.[76] Or by way of a EUR75 million loan for a WtE facility in Lahti, Finland, which will convert municipal waste into heat and power through the large-scale use of waste gasification, gas cleaning and high-efficiency combustion;[77] and through a loan of EUR70 million for the revamping and optimization of an existing 85.2 MW WtE plant at Spittelau located in Vienna.[78] Otherwise, it uses EU funds through the vehicle of a PPP structure – such as in the case of the construction of a WtE facility with an incineration capacity of 210,000 tonnes/year for the municipal waste treatment for the city of Poznan (Poland) and the nine surrounding municipalities.[79]

Furthermore, in 2015 the EIB has launched its Private Finance for Energy Efficiency (PF4EE) initiative, which promotes financing, managed by local banks, for small energy efficiency projects in solid waste management, in particular projects to increase recycling and energy from waste. The instrument is managed by the EIB and funded by the Programme for the Environment and Climate Action (LIFE programme), which has committed EUR80 million to fund the credit risk protection and expert support services. The EIB will leverage this amount, making available a minimum of EUR480 million in long-term financing. More generally, the PF4EE initiative is part of the EIB Operational Plan 2016-2018[80] to fund €71 billion per year over this period, including €17-24 billion a year in higher-risk special activities. These concern, e.g. projects where the EIB has taken on a higher-risk position than the bank's standard operations in order to support the EU's priorities including renewable energy and energy efficiency.

From an operational perspective, the PF4EE instrument – which has already signed operations in the Czech Republic, Spain and France (none of them specifically operating in the area of WtE) – will provide a portfolio-based credit risk protection given by means of a cash-collateral (Risk Sharing Facility), together with long-term financing from the

---

projects; *see* at: <http://balkangreenenergynews.com/new-ebrd-support-renewable-energy-investments-turkey/>.

76 EIB, Press Release, 'EIB Confirms GBP 110 Million Backing for Viridor's Cardiff Energy Recovery Facility (ERF)', 3 March 2016, at: <www.eib.org/infocentre/press/releases/all/2016/2016-058-european-investment-bank-confirms-gbp-110-million-backing-for-viridors-cardiff-energy-recovery-facility-erf.htm>.

77 EIB, Press Release, 'EIB Supports Innovative Waste-to-Energy Project in Lahti, Finland', 3 September 2010, at: <www.eib.org/infocentre/press/releases/all/2010/2010-141-finlande-la-bei-soutient-un-projet-innovant-de-valorisatio n-energetique-des-dechets-a-lahti>.

78 *See* at: <www.eib.org/projects/pipelines/pipeline/20120235>.

79 European PPP Expertise Centre, *Poznan Waste-to-Energy Project, Poland. Using EU Funds in PPPs Case Study*, June 2012, at: <www.eib.org/epec/resources/poznan_case_study_eu_funds_in_ppps_public.pdf>.

80 EIB, *The EIB Group Operational Plan 2016-2018*, Luxembourg, 15 December 2015, at: <www.eib.org/attachments/strategies/cop2016_en.pdf>.

EIB (EIB Loan for Energy Efficiency). This will be available at the EIB's own risk, at competitive rates and with long-maturities and expert support services for the Financial Intermediaries (Expert Support Facility). Financial intermediaries participating in the implementation of the PF4EE instrument will be required to comply with specific criteria. They must, e.g. be a financial institutions of the private sector or operate on the market in a manner comparable to such an institution. Furthermore, they have to be duly authorized to carry out lending or leasing activities according to the applicable legislation and be established and operating in a member state. The selection process will take into account that no more than one financial intermediary per member state can be selected to distribute the PF4EE instrument. Priority is to be given to those financial intermediaries proposing to operate where the needs for energy efficiency investment are the greatest and/or there is an underdevelopment of the use of loan finance for energy efficiency and/or where the capacity to take up loans by final recipients is deemed low.[81]

2.4 CONCLUSION

In this chapter we have focused on current investment policies in the WtE sector developed by major international financial institutions and platforms, namely the World Bank group, the Climate Investment Funds, the World Energy Council, UNEP and the UN Clean Development Mechanism, as well as by some regional organizations – such as the EBRD, the EIB and the ADB – which are particularly involved in this area. Although debt and equity still maintain their leadership as the most effective option for the construction and development of WtE infrastructure projects, particularly in developing countries the application of traditional commercial schemes of WtE financing is often extremely difficult. Indeed, in these countries not only solid waste management, except perhaps in capitals and large cities, is generally given a very low priority, with very limited funds provided, but the problem is even more acute considering that the local taxation system is often inadequately developed and users' willingness to pay for the services that are irregular and ineffective is not high either.

It is in that specific perspective that WtE – being a technology that, at the end of the day, allows us to take waste and turn it into a useable form of energy – has proved to be an attractive, very popular and practical solution in both developed and developing markets to contribute to the resolution of the energy dilemma of ongoing urbanization: an increased rate of generation of all types of waste, on the one hand, and an ever-increasing need for energy, on the other hand. This relationship is clearly illustrated by some data: corresponding to an increase of global waste generation, which will double by 2025 to over 6 million

81 *See, Announcement of the new EIB Product under the Programme for Environment and Climate Action (LIFE) dedicated to Energy Efficiency Investments. The Private Finance for Energy Efficiency (PF4EE) Instrument*, 9 December 2014, at: <www.eib.org/attachments/documents/pf4ee_announcement_en.pdf>.

tonnes of waste per day, WtE market size will grow at a compound annual growth rate of 5.5% from 2016 to almost US$44 billion in 2024. Despite some criticism regarding, above all, its environmental impacts,[82] the "pragmatic" fascination for WtE is a winning factor as is shown in a recent (6 April 2016) declaration of the current European Environment Commissioner, Karmenu Vella, commenting on the European "Circular Economy" package:[83]

> The first objective of the circular economy package is to avoid waste in the first place. But you can never eliminate it all, and you can never recycle all of it. However you can still gain by recovering energy from the non-recyclable materials.

Such considerations, including the traditional difficulty of developing countries to attract foreign private capital, have turned international actors as key player to support financing in the WtE sector, especially if operating under the specific light of a multilateral developmental investment bank (such as the WB Group, the ADB or the EBRD). New initiatives, such as the Climate Investment Funds, the EIB's PF4EE instrument, or the ADB's Clean Energy Financing Partnership Facility, have been developed over the last 10 years in order to help countries to improve energy access, energy security and transition to low-carbon economies through cost-effective investments in technologies and practices, including WtE, that result in GHG mitigation, as is requested, e.g. by the UNFCC and, most recently, in the Paris Agreement. The environmental impact of WtE technologies and their effective contribution to decarbonization clearly represent focal issues in the channelling of funds for the future development of these technologies. The EU Commission has recognized in its recent *Communication on the Role of WtE in the Circular Economy*, that when planning future investments in WtE facilities, more consideration should be given to those processes, such as anaerobic digestion of biodegradable waste, where material recycling is combined with energy recovery.[84] Conversely, the role of waste incineration – currently the predominant WtE option – needs to be redefined to ensure that increases in recycling and reuse are not hampered and that overcapacities for residual waste treatment are averted.[85]

---

82 *See*, for example, J. Morris *et al.*, 'What is the Best Disposal Option for the Leftovers on the Way to Zero Waste', 2013, at: <www.ecocycle.org/files/pdfs/best_disposal_option_for_leftovers_on_the_way_to_Zero_Waste.pdf>. *See, also, World Energy Resources 2016, supra* note 4, pp. 89 ff.

83 *See* at: <www.ciwm-journal.co.uk/considerable-potential-waste-energy-vella/>. For the 'Circular Economy' package, *see* Communication from the Commission, COM(2015) 614 final, Brussels, 2 December 2015.

84 The potential of biodegradable waste coupled with anaerobic digestion processing in a biogas plant is seen in Milan. Since 2014, the city has almost reached 100% collection of food and organic waste, providing an average of 120,000 tonnes of biodegradable waste per year. At full capacity (12.8 MW), the city biogas plant should produce some 35,880 MW for hour of electricity a year, enough to supply 24,000 people, and yield 14,400 tonnes of fertiliser. *See*: <http://european-biogas.eu/wp-content/uploads/2016/03/Milan.pdf>.

85 *Communication on the Role of WtE in the Circular Economy, supra* note 5, p. 11. *See, also*, chapter 1 by Post in this book.

3 WASTE-TO-ENERGY TECHNOLOGIES

Lieve Helsen, Nicholas Agon and Anouk Bosmans

3.1 BACKGROUND

Since the start of the twenty-first century, a slow but noticeable transition in the waste management industry has been in progress. In the past, incineration was the preferred technology to reduce the solid waste in mass (about 70-80%) and in volume (about 80-90%) and to destroy harmful substances in order to prevent threats to human health. Nowadays, waste incineration is almost always combined with energy recovery. Even though the importance of the energy recovery part has increased over time, incineration is not the self-evident choice anymore for newly installed waste treatment capacity, since new projects are also assessing the implementation of more advanced waste conversion technologies. This evolution has been triggered by different environmental and socio-economic pull and push measures that are causing incineration to reach its economic and technological limits.

The public awareness of the detrimental effect of incineration ashes and greenhouse gas emissions on the environment and human health has increased significantly over the last two decades.[1] The evidence and general acceptance of the existence of the hazardous effect of components contained in flue gases (especially dioxins, furans and heavy metals) have led to increasingly stringent legislation on the incineration of waste. Therefore, the aim of the Waste Incineration Directive[2] is to prevent or to reduce as far as possible negative effects on the environment caused by the incineration and co-incineration of waste. In particular, it should reduce pollution caused by emissions into the air, soil, surface water and groundwater, and thus lessen the risks that these pose to human health.

The Waste Framework Directive[3] goes one step further and requires that waste in general needs to be managed without endangering human health or harming the environment. Rather than just focusing on the controlled destruction of waste, clear definitions of waste as a secondary resource are also incorporated in this legislation. The global availability of resources is finite and the rate of depletion of currently known accessible reserves is accelerated by the increased global demand of fossil resources. Concurrently, this makes the extraction of raw materials more difficult and more energy-intensive,

1 H. Apsimon et al., 2009.
2 Directive 2000/76/EC of 4 December 2000, OJ L 332.
3 Directive 2008/98/EC of 19 November 2008, OJ L 312.

resulting in additional environmental burden.[4] Waste is therefore being considered as an interesting secondary resource for material and/or energy purposes.

Waste management has – in accordance with the waste hierarchy as defined in the Waste Framework Directive – evolved to a stronger focus on waste prevention, material recuperation and recycling (e.g. glass, paper, metals) with ambitious targets to be achieved by 2020. This further adds to the diversion of waste from landfills, which was already initiated by the Landfill Directive[5] to significantly reduce the amount of biodegradable material being sent to landfills.

Despite increasing attention paid to waste prevention and sustainability, total municipal solid waste (MSW) generation in Europe has raised from about 150 million tonnes in 1980 (EU25) to more than 260 million tonnes in 2007 (EU28). The total MSW generation for the EU28 has slightly decreased to a little over 240 million tonnes in 2014.[6] Although the waste production in the EU and other Western countries is in a downward trend, the amount of waste generated worldwide is expected to rise exponentially in the coming decades.[7] The world population is rising fast and it is expected to reach 9 billion by 2040.[8] Associated with the growing world population, urbanization in emerging developing countries will increase. It is estimated that 70% of China's population, or approximately 1.0 billion people, will be living in urban areas by 2050.[9] The increasing urban population and changes in lifestyle towards Western standards will increase the volume of generated MSW significantly and will pose a tremendous challenge in terms of waste management.

Today we are already consuming natural resources at a pace one and a half time faster than the Earth can supply,[10] so in order to meet the even higher need for resources of future generations and manage the waste in a sustainable way, a radical change from today's approach is needed.

Important steps are currently being taken to effectuate a transition to a stronger and more circular economy where resources are used in a more sustainable way. In a circular economy, the value of products and materials is maintained for as long as possible. Waste and resource use are minimized, and when a product reaches the end of its life, it is used again to create further value. The European Commission has put forward a Circular Economy Package,[11] which will serve as a guideline and framework for its implementation. Besides focusing on production phase, importance of product design and extended

4 M. Hoel and S. Kverndokk, 1996.
5 Council Directive 1999/31/EC of 26 April 1999, OJ L 182.
6 Eurostat, 2014.
7 T. Karak, R. Bhagat, and P. Bhattacharyya, 2012.
8 United Nations, 2014.
9 J. Feiner, S. Mi, and W. Schmid, 2001.
10 V. Niccolucci et al., 2012.
11 COM/2015/0614.

producer responsibility, a sustainable waste management approach has a crucial role in a successful transition to a circular economy.

Different versions or implementations of the core idea of circular economy have been introduced, such as the concept of Sustainable Materials Management (SMM), introduced by Jones et al.,[12] which comprises the reframing of materials cycles and waste management concepts, targeting closed loop systems. With respect to waste handling, traditional landfilling (i.e. discarding materials in dumps or landfills) cannot be part of SMM as it opposes the idea of a fully closed material cycle. In the novel concept of Enhanced Waste Management (EWM), waste prevention, reuse/recycling and sustainable waste valorization become even more important. Waste valorization is its use as material or the conversion into energy or fuels, with particular focus on environmental indicators and sustainability goals. This means that waste management practices should aim at higher resource recovery potential. Therefore, sustainable waste-to-products (WtP) and waste-to-energy (WtE) technologies are greatly needed. In that context, incineration of waste has made numerous technological advances to maximize the energy recovery from waste. This has also been incentivized by decreasing financial support from governments for WtE facilities. However, combustion is limited in terms of the level of energy recovery that is possible from the feedstock, leading to an increased interest in more innovative, resource-efficient and environmental-friendly conversion technologies for waste.

In this context, the concept of enhanced landfill mining (ELFM) has been introduced by the Flemish ELFM Consortium in 2008. Currently, ELFM is defined as

> the safe conditioning, excavation and integrated valorisation of (historic and/or future) landfilled waste streams as both materials (waste-to-products, WtP) and energy (waste-to-energy, WtE), using innovative transformation technologies and respecting the most stringent social and ecological criteria.

ELFM not only focuses on the most suitable technology to valorize the waste, but also offers the opportunity to select the most suitable moment to valorize certain waste streams, depending for instance on the state of the technology. This is done by integrating landfilling of waste in a sustainable context, where it is no longer considered a final solution, but considered temporary storage places for waste awaiting further treatment or also as future mines for materials. Additionally, the "enhanced" in ELFM incorporates the goal to prevent the emissions of CO_2 and pollutants arising during the energy/material valorization processes. The existing environmental legislation mainly focuses on disposal of waste in landfills and on conventional waste treatment techniques, and is hereby still acting as a barrier to the introduction of innovative waste valorization technologies.

12 P.T. Jones et al., 2013.

The ongoing shift towards more sustainability through valorization of waste as both energy and materials should contribute to improve and adapt the existing policy.[13]

The present chapter reviews WtE technologies using (pre-processed) MSW as input. WtE is the process of recovering energy, in the form of electricity and/or heat, from waste. The focus in this chapter is on available technologies for thermochemical treatment of (calorific) waste streams. The scope is limited to technologies that have been commercially proven in a full-scale plant, or that have at least demonstrated their viability through pilot plant testing. This review summarizes the technological approaches that have been developed, presents some of the basic principles, provides details of some specific processes (more emphasis is put on new advanced technologies) and concludes with a comparison among the different technologies, stressing factors affecting their applicability and operational suitability. The evaluation criteria are based on environmental impact, energy efficiency, material recuperation and system operation (e.g. flexibility in dealing with input variation). Hence, this review constitutes the base for selecting best available technique(s) for energetic valorization of specific calorific waste streams.

3.2 Waste as Energy Feedstock

The term "waste" encompasses a lot of different material streams. Different types of waste include municipal solid waste (MSW), intended as unsorted residual waste (i.e. the waste left downstream of separate collection), solid recovered fuels (SRF) or refuse-derived fuel (RDF), industrial waste and industrial hazardous waste. Table 1 gives an overview of the most relevant types of waste and waste-derived fuels.[14]

Table 1 Different types of waste and waste-derived fuels

Fuel type	Definition
Fuel	Energy carrier intended for energy conversion
Municipal Solid Waste (MSW)	Waste generated by households (may also include similar wastes generated by small businesses and public institutions), e.g. paper, cardboard, metals, textiles, organics (food and garden waste), and wood
Commercial & Industrial Waste (C&IW)	Waste derived from commerce and industry, e.g. packaging, paper, metals, tires, textiles, and biomass

13 *Ibid.*
14 EIONET, 2012; C.J. Lupa et al., 2011; S.T. Wagland et al., 2011; and R. Zevenhoven and L. Saeed, 2003.

Fuel type	Definition
Refuse-Derived Fuel (RDF)	Fuel produced from MSW and/or C&IW that has undergone processing (i.e. separation of recyclables and non-combustible materials, shredding, size reduction, and/or pelletizing), has an input-driven specification
Solid Recovered Fuel (SRF)	Comparable to RDF but considered more homogeneous and less contaminated, is market-driven due to tighter quality specifications
Automotive Shredder Residue (ASR)	Complex mixture of plastics (rigid and foam), rubber, glass, wood, paper, leather, textile, sand plus other dirt, and a significant fraction of metals

Municipal solid waste, the main type of waste considered in this chapter, refers to the materials discarded in urban areas, including predominantly household waste with sometimes the addition of commercial wastes, collected and disposed by the municipalities.[15] MSW is a heterogeneous feedstock containing materials with widely varying sizes, shapes and composition. It contains significant fractions of paper and cardboard, food waste, wood waste, yard trimmings, textile, plastics, etc. MSW is often referred to as a renewable resource; however, the degree of renewability of waste is often a subject of debate. It is closely linked to the biogenic content of the waste. The renewable perception of MSW also varies depending on the context of the discussion. In environmental terms, waste is often considered a renewable energy source.[16] For economic considerations, e.g. when renewable incentives are calculated, MSW is only renewable to the extent that the energy content of the material stream is biogenic.[17] Gohlke[18] determined that generally, more than half of the carbon in MSW is biogenic in origin. However, when the source separation of organic waste, paper and cardboard is carried out successfully, the share of the renewable energy content of MSW may be lower.

A basic prerequisite for waste treatment processes is the adequate characterization of materials contained in the available waste streams. The characteristics of solid waste feedstock are influenced by various factors ranging from storage method (influence on humidity), maturity, sorting policy (differs from country to country) and many more.[19] Characterization data give an indication of the suitability of a specific waste stream for the different valorization options. They are therefore of crucial importance in determining the technical and economic feasibility of available valorization processes. Since the successful implementation of (advanced) thermochemical treatment technologies de-

15 H. Cheng and Y. Hu, 2010.
16 Office of Solid Waste, 2013.
17 Nuclear Energy Information Administration Office of Coal, Electric and Alternate Fuels, 2007.
18 O. Gohlke, 2009.
19 M. Quaghebeur et al., 2013.

pends on the process efficiencies which in turn is dependent on the feed quality. If the MSW is used "as received" as input to WtE processes, this can lead to variable (and even unstable) operating conditions, resulting in quality fluctuations in the end product(s), which have an adverse effect on the process efficiency. Therefore, it is crucial to ensure that the composition and characteristics of MSW fall within the range of the WtE process input requirements. For these reasons, RDF – a processed form of MSW – is often used as input to WtE systems.[20] Refuse is a general term for municipal solid and commercial wastes and the annotations RDF and SRF usually refer to the segregated high calorific fraction of MSW.

In general, the process of converting MSW into RDF consists of shredding, screening, sorting, drying and/or pelletization in order to improve the handling characteristics and homogeneity of the material. These pre-processing steps of MSW are often referred to as mechanical biological treatment (MBT). Metal recovery and drying of the material stream are generally the most common pre-treatment steps. Drying can be done either actively using (excess) process heat or biologically by the reaction heat from aerobic bio-stabilization of the biodegradable fraction of the MSW. The main benefits of converting MSW to RDF are: a higher calorific value (preferred by advanced WtE technologies), more homogeneous physical and chemical compositions, lower pollutant emissions, lower ash content, reduced excess air requirement during combustion and finally, easier storage, handling and transportation.[21] Therefore, a trade-off between the increased costs of producing RDF from MSW and potential cost reductions in system design and operation needs to be found. The effective energy balance of a WtE plant can also change significantly with the decision to pre-process the waste, because of the additional energy use of the MBT processes and inherent system losses. Even though the overall energy balance can suffer from pre-treatment of MSW, other aspects, such as material recovery and landfill diversion, that can benefit from the decision should also be taken into account.

Table 2 shows the composition of MSW and RDF as found in the Phyllis database,[22] both mean values and ranges are given. It is clear from the provided data that the ash content can vary widely, the same is true for the calorific value. The energy content of the waste is not only largely region dependent, but also changes over time. Because of changes in the consumers' habits and increase in the upstream separate collection, the energy content of collected waste has increased over time.[23] The lower heating value (LHV) for the major part of MSW incinerated in the EU augmented from 10.0 to 10.3 MJ/kg from 2001 to 2010.[24]

20 A. Klein, 2002.
21 National Energy Technology Laboratory, 2012.
22 Energy Research Centre of the Netherlands (ECN), 2011.
23 P.S. Calabrò, 2010.
24 D.O. Reimann, 2012.

Table 2 Composition of MSW and RDF: mean values and [min–max]

		MSW	RDF
Water content	wt% wet	34.2 [31.0-38.5]	10.8 [2.9-38.7]
Volatiles	wt% daf[a]	87.1 [87.1]	88.5 [74.6-99.4]
Ash	wt% dry	33.4 [16.6-44.2]	15.8 [7.8-34.5]
LHV[b]	MJ/kg daf	18.7 [12.1-22.5]	22.6 [16.1-29.3]
C	wt% daf	49.5 [33.9-56.8]	54.6 [42.5-68.7]
H	wt% daf	5.60 [1.72-8.46]	8.37 [5.84-15.16]
O	wt% daf	32.4 [22.4-38.5]	34.4 [15.8-43.7]
N	wt% daf	1.33 [0.7-1.95]	0.91 [0.22-2.37]
S	wt% daf	0.51 [0.22-1.40]	0.41 [0.01-1.27]

[a] Dry ash free.
[b] Lower heating value.

3.3 Thermochemical Conversion Methods

The different available technologies and conversion pathways for waste that have been commercially proven in full-scale plants are shown in Figure 1.

Figure 1 Different conversion pathways for waste

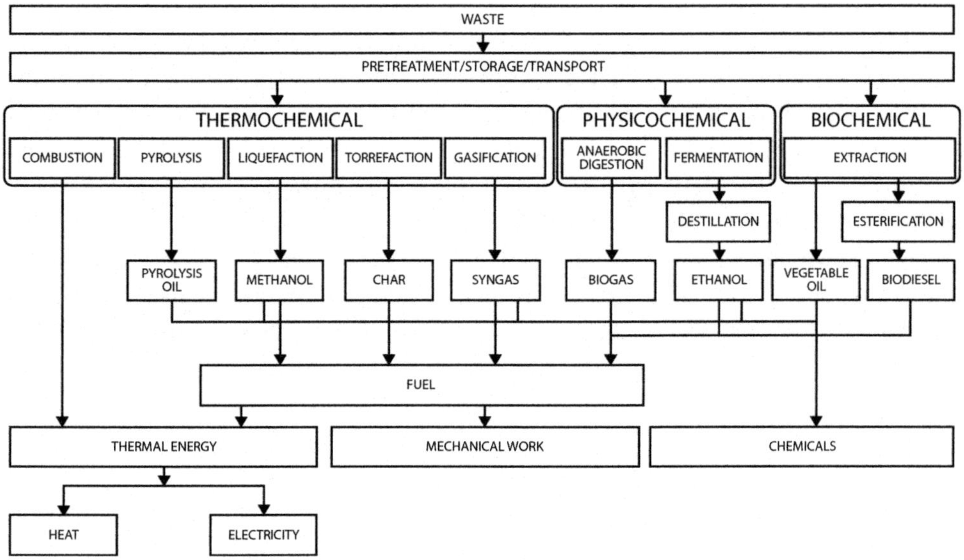

Depending on the type and characteristics (e.g. composition, moisture content, etc.) of the waste, the most suitable treatment option can be selected. According to this overview, the waste after pre-treatment, storage and/or transport can be treated by a range of conversion technologies. The main products obtained from these processes (potentially after an intermediate physical or chemical step) can have different end-uses. The waste-to-resource technologies can be divided into three categories: thermochemical, physico-chemical and biochemical conversion methods. In general, thermochemical treatment, the category that will be discussed in this chapter, refers to any conversion process that takes place at elevated temperatures causing modifications in the chemical structure of the processed material.

The most frequently used approach for energy valorization from waste is combustion, from which the retrieved thermal energy can supply heat and/or generate electricity. Besides incineration, more advanced thermochemical approaches, such as pyrolysis, gasification and plasma-based technologies, have been developed since the 1970s.[25] They involve heating and conversion of the waste in an oxygen-controlled environment to avoid combustion. The primary differences among them relate to heat source, oxygen level, and temperature, from as low as about 300°C for pyrolysis to as high as 20,000°C for plasma gasification. In general, these alternative technologies have been applied to selected waste streams, on a smaller scale than incineration and in specially designed reactors with strictly controlled process conditions. The secondary energy carriers in these pathways can be directly applied as chemical raw materials or utilized as alternative fuel sources. If these are not burned to provide heat and/or electricity, their energy content can be transformed to mechanical work (e.g. to drive a car or fuel a plane). However, when applied to wastes, pyrolysis, gasification and combustion-based processes are often combined, usually on the same site as part of an integrated process. In general, these types of integrated processes recover, in total, the energy value rather than the chemical value of the waste, as would a conventional incinerator do. In a first step, the waste is converted into a secondary energy carrier (a combustible liquid, gas or solid product), while in a second step this secondary energy carrier is burned (in a steam turbine, gas turbine or gas engine) in order to produce heat and/or electricity.

The following sections list the conventional and proven energy recovery options via combustion and discuss and compare the main "advanced" thermochemical conversion technologies (pyrolysis and gasification) and combinations of these technologies. Additional emphasis will be on recent developments in gasification and plasma gasification.

---

25 T. Kolb and H. Seifert, 2002.

3.3.1 Incineration

Basically, incineration is the exothermic oxidation of the combustible organic content in waste and is used as a treatment for a very wide range of wastes. Direct incineration of MSW is also denoted as "mass burn". For effective oxidative combustion, a sufficient oxygen supply is essential and the ratio of the amount of supplied air to the stoichiometric required amount for full oxidation usually ranges from 1.2 to 2.5. During combustion, the majority of the fuel energy is contained as heat in the reaction products (mainly CO_2, H_2O and O_2), also called flue gases. If air is used as the oxidizer, the flue gas also contains nitrogen. Depending on the composition of the material incinerated and on the operating conditions, smaller amounts of CO, hydrogen halides, nitrous oxides, sulphur oxides, heavy metal compounds and tar (e.g. dioxins (PCDD), furans (PCDF) and other volatile organic compounds (VOCs)) are formed or remain.[26] The optimum incineration conditions in which to achieve a good burn out of the gases are a minimum gas phase combustion temperature of 850°C (1,100°C for some hazardous wastes) and a minimum residence time of the flue gases, above this temperature, of 2 seconds after the last incineration air supply.[27] Because of the oxidizing environment and relatively low operating temperatures between 500°C and 1,000°C not all of these harmful components can be broken down and extensive gas cleaning is needed for safe flue gas exhaust. Waste incinerators produce a higher gas volume for the same feed rate in comparison with gasification, pyrolysis and plasma-based systems working under sub-stoichiometric conditions. The gas cleaning equipment scales accordingly. Flue gas cleaning is often an important contributor to overall incineration costs (i.e. 15-35% of the total capital investment). A good control of the combustion process is important in order to reduce pollutant emissions, so improvements in process stability and environmental performance may allow a simplification of downstream flue gas cleaning, which can lead to a significant cost reduction. The performance of the installation benefits greatly from detailed attention to furnace design and air distribution in the reactor volume. These depend strongly on the type of waste that is being treated, with key parameters as the chemical composition, physical and thermal characteristics and their variability. Processes designed for a narrow range of specific inputs can usually be optimized to a larger extent than those that receive wastes with greater variability.

An important concern about waste incineration is air pollution by the flue gases. The legislation on emission standards has therefore continuously evolved to be stricter, enforced by large economic penalties if these standards are being exceeded. The use of modern pollution control technologies has reduced emissions to air to levels at which pollution risks from waste incinerators are now considered to be very low.[28]

26 European Commission, 2006.

27 Directive 2000/76/EC of 4 December 2000, OJ L 332.

28 European Commission, 2006.

The proportions of solid residue (ash, slag, filter dust, other residues from the flue gas cleaning – e.g. calcium or sodium chlorides – and sludge from waste water treatment) vary greatly according to the waste type and detailed process design. The inorganic content and un-reacted carbon (soot) are recovered as fly ash (dust in the flue gas) and bottom ash (i.e. the heavier solid ash). Immobilization of fly ash (generally 1-5% by weight of the input[29]) is required in order to make it environmentally safe for landfill disposal. In MSW incinerators, the bottom ash constitutes approximately 25-30% by weight of the solid waste input. Additional treatment can improve bottom ash characteristics and would allow its use in congregate aggregates and other construction materials. In particular, vitrification receives a lot of attention as a promising technology for the transformation of MSW bottom ash into inert materials. However, since vitrification is an energy-intensive process involving high costs, its use can only be justified if a high-quality product can be fabricated. Research in this field is ongoing.[30]

The three main types of incinerator configurations are mobile/fixed grate (87%), rotary kiln (10%) and fluidized bed (3%).[31] Table 3 summarizes the key features of these three incinerator types.[32] More detailed process descriptions can be found elsewhere.[33]

Table 3 Key features of main incinerator types

	Mobile grate	**Fluidized bed**	**Rotary kiln**
Process description	The grate moves the waste through the various zones of the combustion chamber (tumbling motion)	Lined combustion chamber in the form of a vertical cylinder, the lower section consists of a bed of inert material which is fluidized with air, waste is continuously fed into the fluid sand bed	Cylindrical vessel located on rollers which allow the kiln to rotate/oscillate around its axis, waste is conveyed by gravity
Type of waste	MSW with possible addition of commercial and industrial non-hazardous wastes	Waste with homogeneous particle size distribution (e.g. RDF, sewage sludge)	Any type of waste (incl. hazardous & clinical waste)
Advantages	High level of development	Efficient heat transfer	Suitable for wastes requiring high temperatures and causing severe corrosion
Max. treatment capacity[a]	120 MW	90 MW	30 MW

29 EMIS, 2016.

30 L.M. Schabbach et al., 2012; and L. Machiels et al., 2015.

31 Working Group Energy Recovery (Ramboll), 2012.

32 L. Lombardi, E. Carnevale, and A. Corti, 2015.

33 European Commission, 2006; Limerick, 2005; and UBA, 2001.

	Mobile grate	Fluidized bed	Rotary kiln
Process temperature	850-1,100°C	700-900°C	850-1,300°C

[a] LHV basis in terms of thermal power input per line.

Waste incineration can be an environmentally friendly method if it is combined with energy recovery, control of emissions and an appropriate disposal method for the ultimate waste. Denmark and Sweden have been leaders in using the energy generated from incineration for more than a century. In 2005, waste incineration produced 4.8% of the electricity consumption and 13.7% of the total domestic heat consumption in Denmark.[34] In 2015, the gross electricity production from renewable and non-renewable waste (biomass excluded) in the EU28 amounted to 20,723 and 23,161 GWh, respectively. Together, this only represents 1.4% of the total electricity production in the EU28.[35]

Energy conversion in a WtE plant is typically accomplished in a boiler where the heat content of the flue gases is transferred to liquid water to generate steam. When thermal energy is the only desired end-product, saturated steam is generated. For electricity production or combined heat and power (CHP) production, however, the steam is heated beyond the saturated steam level and superheated steam exits the steam generator. The different configurations with electricity production from waste are briefly discussed in the following sections.

3.3.1.1 Power Production

The production of electricity only is the sole possibility in absence of thermal energy users (industrial plants or district heating) in the proximity of the WtE plant. The superheated steam produced from the recovered heat of the flue gases is supplied to a steam turbine to make a Hirn steam cycle (which is the name commonly used for a Rankine cycle with steam superheating). The average value of electric energy production efficiency, defined as the weighted average specific produced electricity over the weighted average specific energy input (including import), calculated by Reimann,[36] is 20.7%. When compared to efficiencies commonly achieved in conventional fossil fuel-fired power plants (35% for coal, 45% for natural gas), the efficiency for electricity production in a WtE plant appears quite low. A first reason is the small size of WtE plants and the fact that steam turbine efficiency decreases with size.[37] WtE plants are in the range of 100 MW_{LHV} thermal input, while fossil fuel plants range between 1,000 and 2,000 MW_{LHV}. Another reason for the lower efficiencies are conservative steam parameters, because of the limitations of

34 H. Kleis and S. Dalagar, 2007.

35 Eurostat, 2016.

36 D.O. Reimann, 2009.

37 S. Consonni and F. Viaganò, 2012.

the boiler surface materials with respect to temperature stress and corrosion. A detailed description of the above-mentioned reasons and other aspects influencing the electrical efficiency of waste incineration plants can be found in Lombardi et al.[38]

3.3.1.2 CHP or Cogeneration

One of the most obvious means by which the efficiency of WtE plants can be increased is the combined recovery of heat and electricity from the process. CHP is indicated as the best technique for energy recovery[39] and is also found to achieve better performance in life cycle assessment evaluations than electricity production only – provided that the generated heat can be utilized.[40] The improved overall energy recovery also leads to a more positive greenhouse gas balance.[41] In CHP operation, steam can be extracted from the turbine or additional low-temperature steam can be generated after the turbine. In the latter case, the ratio between electricity and heat production is fixed and electricity production is only possible when a constant heat demand exists, since the return fluid from the heat user is the only means for cooling the steam. When steam is used for feeding a district heating network, the amount of generated steam is preferably flexible, to meet the variable heat demand through the seasons. In this case, intermediate steam extraction from a condensing turbine is applied. The European average value for electricity production efficiency (i.e. the weighted average specific produced electricity in co-generation mode over the weighted average specific energy input) is 14.2% and the average heat production efficiency (i.e. the weighted average specific heat produced when CHP is applied over the weighted average specific energy input) is 45.9%.[42]

3.3.1.3 Trigeneration

Trigeneration or combined cold, heat and power (CCHP) is the process by which some of the heat produced by a cogeneration plant is used to generate chilled water for air conditioning or refrigeration. This is achieved by connecting absorption chillers to the network, which are capable of providing a cooling functionality from a temperature difference. In this way, the utilization of surplus heat from the WtE may be valorised also in warmer climates, e.g. by means of district cooling. When the heat from the CHP process is not used for heating, but utilized only for electricity and cooling, this is denoted as combined cold and power (CCP). Taking this process one step further with the addition of systems to purify carbon dioxide from the engine exhaust is termed quadgeneration in industry.

38 L. Lombardi, E. Carnevale, and A. Corti, 2015.
39 European Commission, 2006.
40 A. Damgaard et al., 2010.
41 O. Gohlke, 2009.
42 L. Lombardi, E. Carnevale, and A. Corti, 2015.

The incineration process has undergone rapid technological developments over the last 10-15 years, which have been mainly directed towards reducing emissions. Continuous process development is ongoing, with the sector developing techniques that limit operating costs and at the same time maintaining or improving environmental performance (e.g. optimizing the coating material of the boiler).

3.3.2 *Pyrolysis*

Pyrolysis is the thermal decomposition of organic material at elevated temperatures with limited amounts or in the absence of oxygen. The typical temperature range is from 400°C to 900°C, but usually lower than 700°C. The pyrolysis process is endothermic, so heat needs to be provided to the material. This can be done by partial combustion of the waste material, by direct heat transfer from either the recycled product gas or circulating solids, and by indirect heat transfer from exchange surfaces. Three products are obtained; pyrolysis gas, pyrolysis oil and solid coke (charcoal), whose relative proportions depend very much on the pyrolysis method (slow, fast or flash) and reactor process parameters. The characteristics of the main modes of pyrolysis are summarized in Table 4.[43]

Table 4 Pyrolysis technology variants (RT and HR stand for residence time and heating rate, respectively)

Pyrolysis technology	RT	HR	T_{max} (°C)	Main product
Carbonization	Hours–days	Very low	400	Charcoal
Slow	5-30 min	Low	600	Charcoal, pyrolysis oil, gas
Fast	0.5-5 s	Fairly high	650	Pyrolysis oil
Flash				
- Liquid	<1 s	High	<650	Pyrolysis oil
- Gas	<1 s	High	>650	Chemicals, fuel gas
Ultra	<0.5 s	Very high	1,000	Chemicals, fuel gas
Vacuum	2-30 s	Medium	400	Pyrolysis oil
Hydropyrolysis	<10 s	High	<500	Pyrolysis oil, chemicals
Methanopyrolysis	<10 s	High	>700	Chemicals

Pyrolysis plants for waste treatment usually include the following basic process stages. First, the feed is prepared by grinding, which improves the quality of the waste by increasing the specific surface area and as such promoting the heat transfer to the particles. Second, an

43 A.V. Bridgwater, 2003; and L. Helsen, 2000.

optional separated drying step improves the net calorific value of the raw process gases and increases efficiency of gas–solid reactions within the reactor. Third, during the actual pyrolysis of the material pyrolysis gas is released and a solid carbon-containing residue accumulates, which also contains mineral and metallic portions. Finally, secondary treatment of pyrolysis gas and pyrolysis coke is performed through condensation of the gases for the extraction of energetically usable oil mixtures and/or incineration of gas and coke for the destruction of the organic ingredients and simultaneous utilization of energy.

The calorific values of pyrolysis gas typically lie between 5 and 15 MJ/m^3 based on MSW and between 15 and 30 MJ/m^3 based on RDF.[44] It could be used to power gas engines or gas turbines, which generate electricity more efficiently than conventional steam boilers. Pyrolysis processes have smaller flue gas volumes (after combustion) than incinerators; this reduces the flue gas treatment capital costs to some degree. Pyrolysis processes may offer a number of advantages with respect to material and energy recovery from the feed. It is for example possible to recover (part of) the organic fraction as material/fuel (e.g. as methanol) and the produced char can be recovered for external use, taking into account any necessary pre-treatment (e.g. washing of chlorine). However, pyrolysis of waste produces a large diversity of different substances, which, together with the presence of potential pollutants, can make it (very) difficult to use the pyrolysis products for secondary fuel production or substance recovery from waste.[45]

3.3.3 Gasification

Gasification is the thermochemical conversion of solid, liquid or gaseous carbonaceous material in an oxygen-controlled environment at increased temperatures (usually in the range of 500-1,800°C). The carbon-rich material reacts through partial oxidation to a low or medium value synthesis gas (or syngas) in the presence of an oxidant amount lower than that required for stoichiometric combustion. The resulting syngas consists mainly of CO and H_2, and contains varying amounts of H_2O, CO_2, CH_4, trace amounts of higher hydrocarbons such as ethane and propane, inert gases originating from the gasification agent and various contaminants such as small char particles.[46] The oxygen required for the gasification process can be supplied as free oxygen (pure or as air) or bound in the form of steam, carbon dioxide or a mixture of these gasifying agents.[47] Air gasification produces a low-energy gas (4-7 MJ/Nm^3 LHV), while oxygen gasification produces a medium-energy gas (10-18 MJ/Nm^3 LHV).[48]

44 UBA, 2001.
45 L. Helsen, 2000.
46 A.V. Bridgwater, 2003.
47 A.K. Gupta and W. Cichonski, 2007.
48 L. Helsen, 2000.

3.3.3.1 Process

Gasification is a complex process that includes a number of physical and chemical interactions that occur at temperatures generally higher than 600°C. In the gasifier, the treated material undergoes three major steps: heating and drying, de-volatilization, and gasification.[49] The pyrolysis or de-volatilization process occurs as the carbonaceous particle heats up and thermal cracking of chemical bonds takes place (up to about 700°C). This results in the release of pyrolysis gas (light permanent gases, including CO, CO_2, H_2, CH_4…), tar (a collection of complex organic compounds) and char (the remaining devolatilized residue). The amount, composition and specific features of these fractions depend on the properties of the carbonaceous material, temperature and pressure. The specific gasification reactions are those taking place between the devolatilized solid material (char) and gases excluding oxygen. Two important reactions in this step of converting the solid char to gaseous components are the water–gas reaction with water vapour and the Boudouard reaction with carbon dioxide. Concurrently, the high-temperature environment of the gasification process also promotes the decomposition of tars and heavy hydrocarbons to lighter, desirable hydrocarbons.

The energy required for the thermal cracking and the above-mentioned principal endothermic reactions during the gasification process can be supplied in two different ways. In the majority of gasification configurations, the process is *auto-thermal*, which means that heat is generated by the exothermic (partial) oxidation of part of the feedstock and/or the char (yielding CO and/or CO_2). In an *allothermal* process, energy is supplied by an external source, which can be a heating jacket, heat-carrying bed material or a plasma torch (see further).

3.3.3.2 Products

Starting from a solid feedstock, the gasification process described above generates a gas phase and a solid phase. The obtained syngas contains large amounts of not completely oxidized products, mainly carbon monoxide and hydrogen, that still have a considerable calorific value as opposed to the product gas resulting from conventional direct combustion. That chemical energy can be valorized by deploying the syngas as a feedstock for the chemical industry (through some reforming processes), as a source of environmentally benign fuels or for efficient production of electricity and/or heat.[50] The production of syngas as an intermediate product allows the energetic value of the waste to be recovered by a wide range of applications. Since the focus of this chapter is the transformation of waste to energy, the chemical valorization options will not be discussed in detail, but can be found elsewhere.[51]

49 S.R. Turns, 2000.

50 UBA, 2001.

51 J. van den Loosdrecht and J.W. Niemantsverdriet, 2013.

The syngas at the output of the gasifier is contaminated by various unwanted components such as particulate matter, condensable hydrocarbons (i.e. tars), sulphur compounds, nitrogen compounds, alkali metals (primarily potassium and sodium), heavy metals and acid gases (hydrogen chloride and carbon dioxide). These contaminants can have a damaging effect on both equipment and environment, and need to be reduced to the level required by the downstream syngas conversion technology. Syngas cleaning is therefore a crucial step between gas production in the gasifier and gas utilization.[52] A very comprehensive review of syngas cleaning techniques is made by Woolcock and Brown.[53]

Inorganic matter comprising metals, inerts (such as rocks) and ash remains as a solid by-product from gasification. Metals are to a great extent recovered and reused, but the ash residues are a challenging aspect of waste gasification. They can give rise to significant environmental issues due to the large quantities produced and the potential for leaching of toxic substances (e.g. heavy metals such as lead and arsenic) into the soil and groundwater at disposal sites.

The ash is retrieved either in a fine granulated form or as a glass-like slag. The latter physical structure is achieved when the temperatures in the gasifier are higher than the fusion point of ash. These types of gasifiers are termed slagging or agglomerating gasifiers. Most of the inorganic matter, i.e. non-volatile metals and other mineral compounds, are bound together in molten form and quenched into a solid matrix by a water bath, or by natural heat loss at the bottom of the gasifier. When ash is collected as a powder-like material after gasification, it can be conditioned and used as an aggregate material, melted in dedicated ash vitrification facilities, or disposed of in a landfill.

Vitreous slag is more preferable than ash, because of its habit of encapsulating toxic species (such as heavy metals) into a stable, non-leachable material. Leachability data obtained from different gasifiers unequivocally show that gasifier slag is highly non-leachable, and can be classified as non-hazardous. Because of its particular properties and non-hazardous, non-toxic nature, slag is relatively easily marketed as a by-product for multiple advantageous uses. In addition to the use as an aggregate material, it has potential for use in other applications where technological specifications require some combination of strength and particle size, such as pipe bedding and fill materials. Furthermore, it can be a raw material for more technically demanding applications, such as insulation wool manufacture. The molten state of the slag at the point of generation is then spun on high-speed spinning wheels generating a mass of fine, intertwined fire-resistant fibres. Other products, such as alumino-silicate geopolymers, would utilize the alkali-rich activated alumino-silicate content of the slag. Geopolymer binder technology based on the by-product of a gasification system can offer a "green" alternative to Portland cement.[54]

52 P.V. Aravind and W. de Jong, 2012.

53 P.J. Woolcock and R.C. Brown, 2013.

54 R Taylor, C. Chapman, and A. Faraz, 2013.

The main advantage of gasification processes over combustion is the much smaller volume of gas exiting the process (up to a factor 10 by using pure O_2). This significantly reduces the installed gas cleaning capacity and the associated parasitic load and (operating and investment) costs. Consequently, the waste water flows from synthesis gas cleaning are smaller. Other advantages are owed to the reducing atmosphere in the gasifier, which limits the emission of dioxins, furans and NO_x and improves the quality of the solid residue, particularly with respect to metals, which can be collected at the bottom in non-oxidized form.[55]

3.3.3.3 Material Input

Gasification is a well-established process that has been practice-tested, primarily with coal (±49%), and petcoke and petroleum residues (±40%). The remaining 11% of feedstock consists of a wide range of carbonaceous material streams such as natural gas, biomass and municipal solid waste.[56] The operational experience with the fossil-based feedstocks has revealed how to overcome specific issues related to a certain feedstock. The high sulphur content in petcoke, for example, requires the gasification to be operated at the highest gasification temperature and additional arrangements are needed for the removal of sulphur from the syngas. Moreover, the high ash content in coal can create various operational problems in the gasification system.

For a given gasifier the admissible range of feedstock properties is rather narrow (much more narrow than for a combustor). The chemistry and the fluid-dynamics of gasification are very sensitive to variations of feedstock composition, moisture, ash content, particle size, density, reactivity, etc.[57] Variation in compositions/properties of feedstocks can result in variable operating conditions and consequently in quality fluctuations in the produced gas.[58] It was for instance shown that a difference in the amount of ash between two coal samples had an effect on syngas composition, heating value and cold gas efficiency.[59] It is clear that the performance of the gasifier (efficiency and syngas quality) and the quality of the feedstock are strongly linked. In general, high sulphur and ash content increases syngas cleaning cost and reduces the gasifier lifetime, while low carbon content reduced the LHV of syngas.

The highly heterogeneous nature of municipal solid waste causes significant challenges with respect to the gasification process and feedstock handling. Specific problems to the waste feedstock are: large variability of composition, high concentrations of contaminants, unfavourable chemical properties (high moisture content, low LHV, high ash content) and unfavourable physical properties (irregular particle size, low density). More-

55 U. Arena, 2012.
56 A. Valero and S. Uson, 2006.
57 S. Consonni and F. Viaganò, 2012.
58 P. Mondal, G.S. Dang, and M.O. Garg, 2011.
59 B. Kar, 2006.

over, this variation is time-dependent. Several gasification processes are available or being developed, which are in principle suited for the treatment of MSW, certain hazardous wastes and dried sewage sludge. Good operation of the gasification reactor (i.e. high conversion efficiencies and minimal tar formation) requires that the nature (size, consistency) of the waste input remains within certain predefined limits. Typically, this requires dedicated pre-treatment of MSW to recover the combustible fraction of the waste. The more homogeneous physical and chemical properties of RDF (especially higher LHV and lower moisture content) result in better feed handling, storage and transportation. Keep in mind that pre-treatment by the MBT processes increases the cost of the operation and that their additional energy use should be included in the overall energy balance of the system.

In the last 30 years, a lot of industrial gasification projects have been built, amounting to a total capacity of more than 104 $MW_{th}$of syngas heating value in 2013, with another 63.4 MW_{th} in construction and 84 MW_{th} planned.[60] These installations not only include coal and oil-based units, but also a considerable number of plants fed by biomass and waste feedstocks. Where the United States is faced with proposed regulations from the Environmental Protection Agency that would cease the construction of new coal-fired facilities and retire many existing coal-fired power plants, China plans the construction of 50 new coal gasification plants, aiming to produce 50 billion cubic meters of gas from coal by 2020 to cope with its exponential energy demand.[61] Even though the operating experience and the scale of commercial applications for gasification is still more limited for biomass than for coal, some large-scale biomass gasification installations have been successfully brought into operation.

Besides the feedstock characteristics, the properties of syngas also depend on the type of gasifier used, and the operating conditions, which will now be discussed.

3.3.3.4 Operating Parameters

The gasification process is more complex than combustion and there are a series of operating parameters that are crucial for the system's proper operation.[62] Some of these operating conditions, influencing the quality and performance of the gasification process, are briefly discussed in the following.

Process temperature is one of the key parameters of the overall gasification process. High operating temperatures promote carbon conversion (resulting in low char and tar content in the product streams) and positively affects the conversion of NH_3 to molecular nitrogen. The temperature level should nevertheless be carefully controlled, since too high operating temperatures can favour the formation of unwanted tar components

60 C. Higman, 2013.
61 Institute for Energy Research, 2014.
62 J.A. Ruiz et al., 2013.

(such as naphthalene, BTEX or pyrene), decrease the energy efficiency and increase the risk of ash sintering and agglomeration.

Since the oxygen content in the waste is generally not sufficient to ensure complete carbon conversion to syngas, an external source of oxygen needs to be added. Oxidizing media can be air, oxygen-enriched air, pure oxygen, steam or carbon dioxide. Air-blown gasification is the simplest and most economical way to convert waste into a low-calorific syngas. The low energy content of the produced gas (LHV between 4 and 7 MJ/Nm^3) is naturally caused by the dilution with atmospheric nitrogen, inherently present in air. For enhanced chemical conversion to a medium-calorific syngas (characterized by a heating value in the range of 10-20 MJ/Nm^3), inert gases such as nitrogen need to be removed from the gasifying agent. Some processes are operated with oxygen-enriched air, i.e. a mixture of nitrogen and oxygen with an oxygen content up to 50%. This produces a high-calorific gas without the consumption of expensive pure oxygen. Steam gasification can also provide an attractive alternative for the production of medium-calorific syngas. This process is especially attractive when increased hydrogen content in the syngas is desired. No exothermic reactions are involved in steam gasification, so an external source of energy for the endothermic gasification reactions is needed. This also holds when CO_2 is used as an oxidizing medium (either exclusively or in combination with oxygen or steam). The combination of the above-mentioned gasifying agents is also used. The mixture of oxygen and water vapour as oxidizing medium in the gasification process, called oxy-steam gasification, is most performant and is being used more frequently.[63]

Other reactor conditions affecting the gasification conversion efficiency and syngas quality are residence time, equivalence ratio (i.e. the ratio of the actually used amount of oxygen compared to the stoichiometric required amount for complete oxidation) and operating pressure. About the latter, the main advantage of pressurized gasification is the reduced equipment size. In addition, for certain downstream syngas applications, the necessary compression step can be omitted when gasifying under elevated pressures.

3.3.3.5 Configurations

The vast majority of all gasification units ever built were designed to process either coal or petroleum residue. Only in the last couple of decades has the well-established gasification process been reapplied to biomass (and MSW) as well. The majority of the gasifier configurations are therefore mainly designed and optimized for the conversion of coal (and to lesser extent, of biomass) to syngas for the production of electricity. The waste material must be finely granulated for utilization in gasifiers. Thus pre-treatment is necessary, especially for MSW. Hazardous wastes may be gasified directly if they are liquid or finely granulated.

63 K. Sandeep and S. Dasappa, 2014.

In general, gasifiers are categorized based on the type of bed and type of flow. The fixed bed, fluidized bed, entrained flow, rotary kiln, multi-stage gasification and plasma gasification will be described in the following subsections, as the main gasification configurations encountered in WtE applications. Figure 2 shows some of them.[64]

Figure 2 Different types of gasifiers

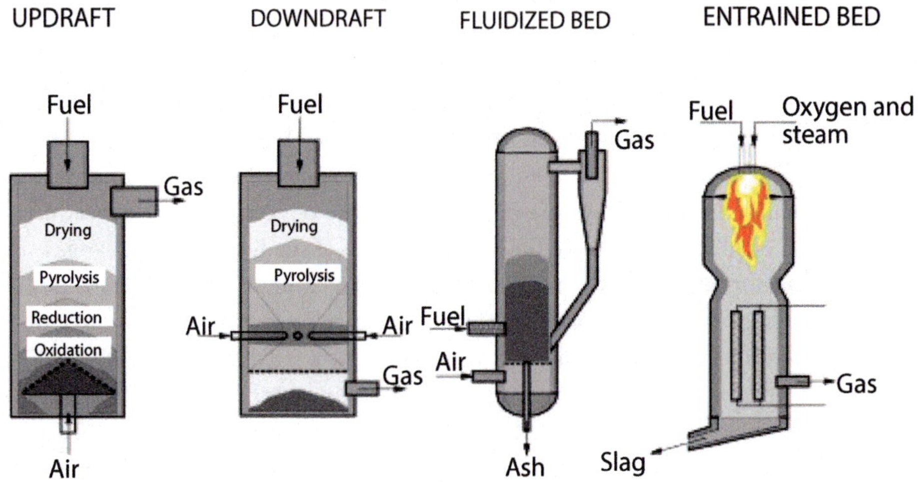

3.3.3.5.1 Fixed Bed

In fixed bed gasifiers, the volume of the reactor is almost completely filled by the waste bed. The bed moves down slowly as a plug flow, passing through different reaction zones in the reactor. This type of configuration can be further differentiated in the way the reagents and the gasifying agent come into contact with each other. In an *updraft or counter-current* reactor, the waste feeding system is at the top of the gasifier and the oxidant intake is at the bottom, so that the product gas moves upwards. The counter-current contact mode contributes to an efficient internal heat exchange, which reduces the syngas exit-temperature to about 200-300°C, albeit with a relatively high tar content (up to 100 g/Nm^3). In *downdraft or co-current* reactors, the material is fed from the top and drops downwards, while the gasifying agent is injected from the side and mingles with the pyrolysis products. The solids and product gas then move downward in parallel streams. The syngas exits from the bottom at higher temperature (around 800°C), with lower tar level compared to an updraft configuration. Of the two, the downdraft configuration is mostly applied in industrial plants. Two commercial examples are: Vivera's Thermoselect technology (licensed to Interstate Waste Technologies) and the Direct

64 P. Mondal, G.S. Dang, and M.O. Garg, 2011.

Melting System (DMS) from Nippon Steel. In the DMS process, no pre-treatment of MSW is required such that it can be directly charged into the gasifier from the top together with coke and limestone, which function as a reducing agent and a viscosity regulator, respectively. With oxygen-rich air blown at the bottom of the furnace, the MSW passes through the typical stages. It is first gradually dried and preheated in the upper section, the combustible waste is then thermally decomposed in the second zone where most of the syngas is produced. The syngas is discharged from the top of the gasifier, but the incombustible waste descends to the combustion and melting zone (1,000-1,800°C) at the bottom and is melted with the heat generated by coke burning.[65]

3.3.3.5.2 Fluidized Bed

In the category of fluidized bed-type gasifiers, bubbling or circulating fluidization is possible. In a *bubbling fluidized bed* (BFB) gasifier, the flow of gaseous oxidant is blown upwards through a distributor plate and permeates a bed of inert material (typically silica sand or olivine) located at the gasifier bottom, which contains the waste to be treated.[66] The resulting fluid-like state achieves an intense mixing between gas and solids allowing very high heat and mass transfer rates. The reactors are characterized by a uniform temperature distribution, which is advantageous for process stability and also decreases NO_x emissions. A bubbling fluidized bed gasifier is used by Enerkem in one of the most exciting and pioneering waste-to-resources facilities in the world. At their biorefinery facility in Edmonton, MSW is converted into methanol using gasification as the intermediary thermochemical conversion method. The process is operated under relatively low temperatures of 700-750°C and moderate pressures of about 2 bar and oxygen and steam are used as fluidizing gases and gasification agents. The installation is capable of handling post-sorted MSW (after recycling and composting) with no need to pelletize it. It is designed to process 100,000 dry metric tonnes of RDF per year, converting it to about 38 million litres of methanol per year.[67]

EBARA Corp. and UBE Industries Ltd. developed a two-stage pressurized gasification and slagging process (EUP) comprising a low-temperature and high-temperature gasification reactor, both operating under elevated pressure (7-8 bar). The technology is applied in Japan to generate synthesis gas from pre-processed waste plastics, while also recycling metals and glass granulate. In the first process stage, gasification takes place in an oxy-steam fluidized bed reactor at low temperatures (600-800°C) to avoid melting of metals like aluminium. This allows a high recovery rate of metals at the bottom of the gasifier in non-oxidized and thus readily marketable form, making the technology particularly attractive for the treatment of waste streams with high metal content (e.g. automotive shredder residue (ASR)). On the other hand, low temperatures slow down gasifi-

65 N. Tanigaki, 2015.
66 U. Arena, 2012.
67 M. Grant, 2016.

cation reactions thus the gas flow leaving the gasification reactor still carries a high load of combustible material. The second stage consists of a cyclonic high-temperature gasifier designed to handle flows with a significant content of solids above their melting point (1,300-1,500°C). Molten ashes are collected in the quenching bath at the bottom of the cyclonic reactor where they solidify into a totally inert, vitrified granular slag. Presently, a number of plants are operating or under construction in Japan.[68]

Circulating fluidized bed (CFB) gasifiers have higher flow rates which carry most of the bed particulates and un-gasified material to an attached cyclone separator, from which the solids are re-circulated to the gasifier bed. This is one of the preferred technologies for large-scale biomass gasification plants.[69]

This type of gasification system is for example used in the world's largest RDF gasification power plant. Since 2011, Lahti Energy has operated the largest and most efficient waste gasification power plant (in Lahti, Finland), processing 250,000 tonnes of RDF per year and producing 50 MW of electricity and 90 MW of district heat from 160 MW of solid waste fuel power. The installation consists of two 80 MW_{fuel} CFB gasifiers using air as an oxidizer and operating at atmospheric pressure and temperatures of approximately 900°C.[70]

3.3.3.5.3 Entrained Flow

Entrained-flow reactors are typically operated at high pressure, in the range of 20-70 bar, and high temperatures, in the range of 1,300-1,500°C. This configuration relies on the injection of very fine feed particles with diameter smaller than 1 mm, possibly mixed with water into a slurry. This has been successfully achieved for mixed plastic waste, but may pose a problem for the use of other waste types as feedstock. The high operating temperature and pressure imply a very low tar and methane content, high throughput and the possible production of a vitrified slag (which is the case for a slagging gasifier).

3.3.3.5.4 Rotary Kiln

The rotary kiln reactor is composed of a slightly tilted (1-3%) steel cylindrical shell, that slowly rotates around its own axis at about 1.5 rpm. The rotation of the drum ensures the gas–solid contact between the waste and the gasification medium, by continuously stirring. The conditions of heat and mass transfer are not very effective, resulting in a long residence time and low syngas quality. This issue may be solved by installing barriers inside the drum to enhance the contact, dividing it into different chambers.

A complete overview of the differences between the different gasifiers and the (dis)-advantages of each type has been presented by Warnecke.[71] Other variations on gasification

68 EBARA Environmental Plant CO. LTD., 2016; and C. Steiner et al., 2002.
69 J.A. Ruiz et al., 2013.
70 M. Bolhar-NordenKampf and J. Isaksson, 2014.
71 R. Warnecke, 2000.

processes have been tried and are being developed, for a variety of waste streams. Examples can be found elsewhere.[72] Table 5 summarizes the key features of each gasifier type.

Table 5 Process characteristics of the three main gasifier types for waste treatment (based on Bridgwater[73])

	Fixed bed	Fluidized bed	Entrained flow
Process description	- downdraft: solid and gas move down - updraft: solid moves down, gas moves up	- bubbling: low gas velocity, inert material stays in reactor - circulating: inert material is elutriated, separated and recirculated	- usually no inert solid, high gas velocity - can be run as cyclonic reactor
Process temperature	1,000°C	800-850°C	1,200-1,500°C
Remarks	- simple and robust construction - finely granulated feedstock required - downdraft: low moisture fuels required, low tar content in product gas - updraft: low exit gas temperature, high levels of tar in product gas	- greater tolerance to particle size range than fixed beds - moderate tar levels in product gas - bubbling: tolerates variations in fuel quality - circulating: operation more difficult than fixed beds	- finely granulated feedstock required - low tar and methane content in product gas - potential slagging of ash

3.3.3.5.5 Multi-stage Gasification

In recent years, another distinction in the design between gasification systems, mainly for waste, can be made. The classification is made based on the level of integration of the different gasification sub-processes, i.e. drying, pyrolysis and gasification. The gasifier configurations described above, are integrated concepts where these sub-processes are occurring in the same reactor volume, but when the drying, pyrolysis and/or gasification zone are physically separated, the concept is called *multi-stage gasification*. This enables waste conversion into syngas to take place under optimized operational settings for each individual step. The main objective behind this concept is to obtain a high-quality clean syngas with a lower tar content. Improved efficiency and larger throughput have already been proven for biomass gasification using this multi-zoned reactor set-up.[74] One exam-

72 European Commission, 2006; A.V. Bridgwater, 2003; A.V. Bridgwater, 1995; and T. Malkow, 2004.
73 A.V. Bridgwater, 1995.
74 S. Heidenreich and P.U. Foscolo, 2015.

ple of the multi-stage gasification is the WoodRoll technology by Cortus Energy.[75] In this system, biomass first passes through the dryer. Subsequently, the dried biomass is thermally decomposed in the pyrolysis chamber into solid char and pyrolysis gas. The pyrolysis gas, containing tars and other volatile components, is carried out of the system. The energy released from combustion of part of the pyrolysis gas is used to deliver the heat to the drying step. The other part of the pyrolysis gas is combusted in multiple set single ended recuperative burners inside the gasifier module. The gasifier is fed with pyrolysis char and steam. The heat for the gasification process is indirectly supplied from the radiating tubes of the burners. The heat carried by the syngas exiting the gasifier is recovered in a heat exchanger for evaporating incoming water to steam (used as gasifying agent). In this way, pure synthesis gas is produced without nitrogen.

A highly technological approach for multi-stage gasification of waste is introduced by GreenE.[76] After a drying step, pyrolysis of the material takes place in a six-chamber rotary kiln pyrolysis reactor. The resulting gas fraction (termed pyrogas) is sent to a cracking chamber where undesirable condensable tar components are further broken down at elevated temperatures. The residual solid (char) fraction on the other hand is further converted to syngas in a separate gasification chamber with the addition of steam. The mixture of produced syngas and cracked pyrogas passes through a heat-exchanging step and a cleaning section before it is introduced in a cogeneration engine. The exhaust gases from the combustion of the gas mixture are fed back and further heated in the heat-exchanging step by the heat of the product gas mixture. The combustion gases circulate through the pyrolysis reactor to provide energy for the pyrolysis process. Residual thermal energy is then recovered with a boiler to produce steam for gasification that is further preheated by the same previously mentioned heat-exchanging step before it is fed to the gasifier.

Another company that has developed a process that allows waste conversion into a clean syngas is Xylowatt with their NOTAR© technology.[77] The principle also relies on the physical separation of the different reaction zones by dividing the gasification process into a pyrolysis, combustion and reduction step. At the top of the vertical orientation of the system, partial combustion of the raw feedstock provides the necessary heat for complete pyrolysis of the fuel. In the second step, the gaseous phase comprising pyrolysis gases and tar is burned in the combustion zone, destroying any remaining tars in the process. The injected air does not come into contact with the char, avoiding the creation of clinkers. The tar-free char then further reacts with the flue gases from the oxidation zone (CO_2 and H_2O) in the reduction zone to produce syngas. The integration of three

75 M.N. Folkelid, 2016.
76 GreenE, 2016.
77 A. Bacq et al., 2016.

separate compartments assembled in a compact way with the NOTAR technology offers accurate control of the operating parameters and is able to produce a clean syngas.

3.3.3.5.6 Plasma Gasification

Plasma gasification is a particular type of configuration in which the gasification process is combined with a thermal plasma. In order to describe how a plasma element can have an added value to conventional gasification, it is first necessary to give a brief explanation on plasma.

Plasma is described as the fourth state of matter, next to solid, liquid and gas states. In forming a plasma, the molecules of a gas get sufficient energy and dissociate. If the thermal motion is strong enough, more and more atoms will be split into electrons and ions until the gas becomes ionized and is transformed into a plasma. The definition of plasma is, in fact, a quasi-neutral ensemble of neutral, and positively and negatively charged particles featuring collective behaviour. The presence of charged gaseous species makes the plasma highly reactive and causes it to behave significantly differently from other gases, solids or liquids. Plasmas can be categorized into different categories depending on electron temperatures and electron densities, but for the purpose of plasma gasification, the focus is on thermal plasmas. The most common method of supplying energy to a neutral gas to form a thermal plasma (as described above) is by means of an electric field. These are called arc-discharge plasmas since the conductive path of charged species in a plasma between the electrodes can cause an electrical breakdown and sustain an electric arc (similar to lightning). The device that incorporates the plasma forming components and stabilizes (by gas or water) the electrical discharge, allowing a continuous and directed flow of plasma, is called a plasma torch. The primary electricity sources used to generate the electric field are predominantly direct current (DC) and alternating current (AC). Further distinctions between plasma torches can be made based on the operating mode (i.e. non-transferred versus transferred, depending on whether the arc is created between two electrodes within the housing of the torch or not), the plasma-forming gas (Ar, H_2O, air, CO_2,...), electrode types, arc stabilizing mechanisms, etc. These fall out of the scope of this chapter and will not be discussed in detail.[78] DC plasma torches are most frequently used for plasma gasification. They are typically operated at atmospheric pressure and the plasma flow is characterized by high enthalpy content and high temperatures, typically from 2,000°C up to 20,000°C.

These characteristics make thermal plasma suitable for a diversity of industrial applications, such as plasma spraying, plasma cutting and plasma welding techniques. Besides these well-established manufacturing applications, thermal plasma has also triggered the interest for its use in waste materials treatment. The application of plasma-based systems to waste management is a relatively new concept, since it was initially used to melt or

78 P. Fauchais and A. Vardelle, 1997.

immobilize solid materials such as ash or metals, encapsulating the harmful elements in a non-leachable vitrified matrix, making them safe for disposal. This has been further developed over the years, to the point where plasma-based treatment has become important for the destruction of hazardous liquids and gases and is a thoroughly tested, commercially proven technology. The PLASCON process (developed by CSIRO and SRL Plasma Ltd. in Australia, and now owned by DoloMatrix International Ltd.) applies plasma pyrolysis to fluid wastes containing halogenated hydrocarbons, CFCs, HFCs, PCBs and other harmful components. Based on the most recent information, 10 plants are operating in Australia, Japan, the United States and Mexico.[79] There exist other relatively small-scale plasma pyrolysis installations for treating polymers,[80] medical waste[81] and low-level radioactive waste,[82] but no information was found on industrial plasma pyrolysis facilities processing MSW or RDF. Another example of plasma-based treatment of solid wastes that require decontamination as well as volume reduction and immobilization of inorganic contaminants, is the Europlasma plant. Europlasma uses plasma reactors with non-transferred arc torches for incinerator residue compaction, the waste is heated directly with plasma jets. The plasma direct melting reactor installed in Cenon (France) processes up to 10 tonnes of fly ash per day, a larger installation in Shimonoseki (Japan) processes daily 42 tonnes of fly ash and bottom ash.

The field of application for thermal plasmas in materials treatment further expanded from hazardous wastes to MSW. It was recognized that plasma could offer a number of advantages to gasification processes.[83] Thermal plasmas can be combined with gasification in two ways: (1) it can serve as the external heat source for an allothermal gasification process, or (2) it can be used as a downstream processing step after the gasification process.

The former set-up, where the thermal plasma torch is directly incorporated in the gasification reactor and supplies the energy for the gasification reactions, is called *single-stage plasma gasification*. As energy for the process is supplied by the plasma power, chemical reactions with the treated material and its devolatilization products are not the primary source of energy. In this way, the injected energy can be controlled independently of the heating value of the treated material. Also, the process temperature can be set independently from fluctuations in feed quality and supply of air/oxygen/steam required to gasify the feed.[84] Plasma allows a better process control than other energy carriers in the allothermal gasification process. Whereas an upper temperature limit of 2,000°C can be achieved by burning fossil fuels, electrically generated thermal plasmas

79 J. Heberlein and A.B. Murphy, 2008.
80 R.R. Guddeti, R. Knight, and E.D. Grossmann, 2000.
81 S.K. Nema and K.S. Ganeshprasad, 2002.
82 B.Y. Min et al., 2007.
83 J. Heberlein and A.B. Murphy, 2008.
84 B. Lemmens et al., 2007.

can reach core temperatures of 20,000°C or more.[85] Since electricity is used as the energy source, heat generation is decoupled from process chemistry, which increases process controllability and flexibility. Additionally, the enthalpy provided by the plasma can easily be adjusted by tuning the electrical power supplied to the system. At the elevated temperatures in the reactor volume, the inorganic constituents of the treated material can reach their melting point and can be reduced to a non-hazardous melt with a very low unburnt organic residue content. That is, however, not the primary goal of the plasma source in single-stage plasma gasification.

Westinghouse (Madison, USA) has developed several generations of plasma gasification reactors that can be labelled single-stage systems (see Figure 3). Their design includes a non-transferred plasma torch to provide part of the heat required for waste processing, while the remainder of the process energy is provided by the calorific value of the waste and/or by the addition of coke. The Westinghouse plasma reactor is a plasma-fired furnace containing the waste and coke (about 4 wt% of the total load). Plasma heating of a fraction of the air reduces the amount of coke and air needed to generate the high temperatures in the furnace. The waste composition determines how much of the incoming air needs to be plasma heated. More detailed information about the Westinghouse process and its implementation can be found elsewhere.[86]

Figure 3 Gasifier evolution of Westinghouse Plasma Corporation[87]

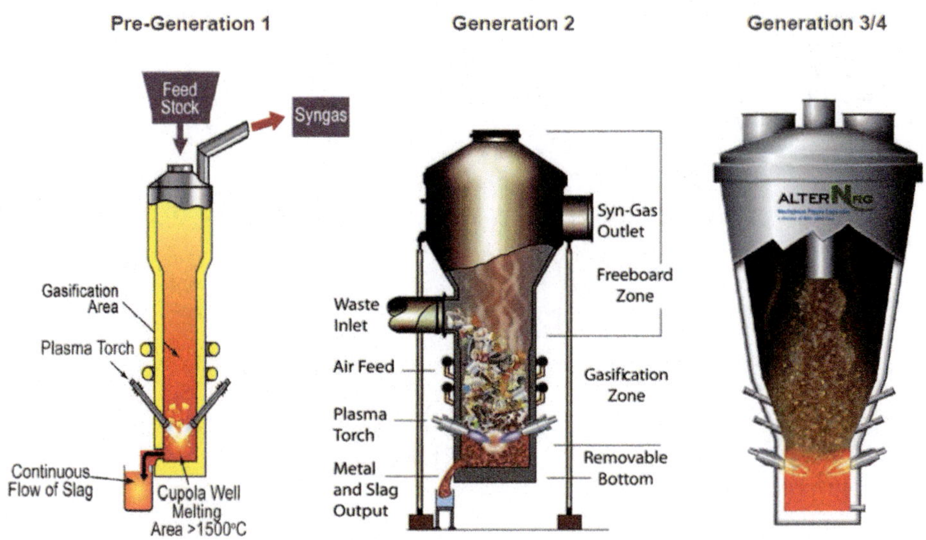

85 E. Gomez et al., 2009.
86 Juniper Consultancy Services Limited, 2008.
87 Westinghouse Plasma Corporation, 2012.

In a *dual-stage plasma gasification* system (the second plasma gasification set-up), contrary to single-stage, the plasma actions happen in a separate step, following the gasification process. In this configuration, thermal plasma processing can act as a syngas cleaning step and/or ash immobilization step. The syngas cleaning action of plasma is achieved by destroying most of the unwanted contaminants in the gas stream. The high-temperature, reactive environment accelerates the kinetics and improves the thermal decomposition of hazardous organic compounds (such as tar) into desirable gas species. This results in a synthesis gas that is cleaner compared to conventional gasification processes. Alternatively (or additionally), the high-temperature plasma plume can be directed to the residual solid fraction from the gasification process. The inorganic components (ash and metals) are then melted and converted into a dense, inert, non-leaching vitrified slag, after quenching.

Figure 4 Illustration of the Europlasma gasification system including Turboplasma unit[88]

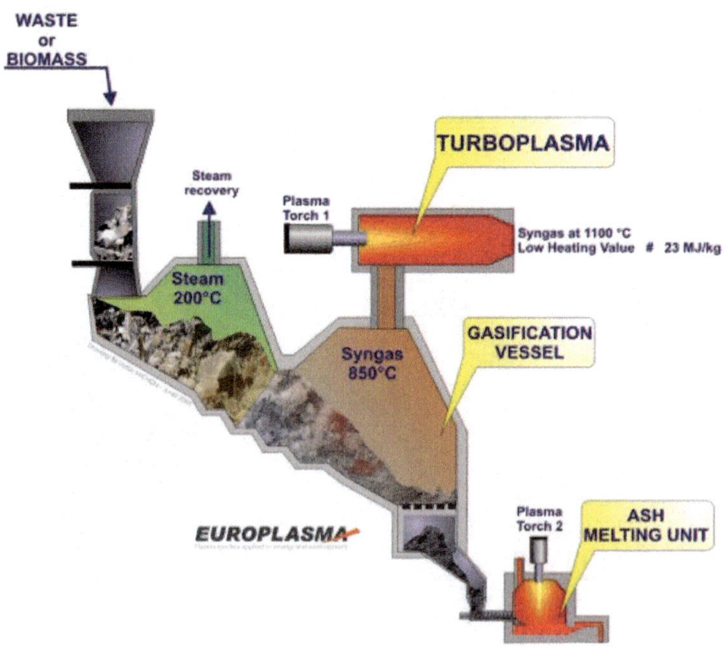

Europlasma has built a 12 MW gasification plant for solid waste treatment in Morcenx (France). The design includes two plasma torches, a first one to refine the raw synthesis gas produced during gasification, and a second one to vitrify metals and minerals (see

88 N. Agon, 2015.

Figure 4). It is not clear, however, to which extent the plasma treatment (patented as TurboPlasma©) replaces the different synthesis gas cleaning stages. Tetronics developed a similar process (the Gasplasma© process), which combines fluidized bed gasification with plasma cleaning of the resulting hydrogen-rich syngas. The syngas production process comprises two steps; first, in a fluidized bed gasifier VOCs and carbon are converted into a crude syngas using a fraction of the thermal energy of the waste and next, a plasma converter provides the high-temperature environment for converting residual tars and chars, allowing further cleaning of the syngas and vitrification of the ash into a non-leaching slag. Rigorous pilot plant trials are being performed by Advanced Plasma Power (APP) on RDF from a different number of sources (including MSW excavated from a landfill). The first results are promising and show that this technology offers great potential for waste valorisation. More detailed information on the Gasplasma© process can be found elsewhere.[89]

Both single- and dual-stage plasma gasification systems have benefits and disadvantages. For selecting the most optimal configuration, the waste stream is one of the most important parameters – as it is in all instances of decision-making for waste treatment processes. Since the single-stage gasification process does not rely on the energy content of the treated material, it can handle waste with both low and high calorific value. The energy contained in a plasma allows the use of low-energy fuels, such as household and industrial waste that often cannot sustain their own gasification without additional fuel. On the other hand, the allothermal plasma process can maximally transfer the (medium to high) energy value of waste to synthesis gas. The composition of the syngas is desirable with low levels of CO_2 and inert gases. However, the energy use to supply the plasma power is high, which leads to an overall low energy efficiency. Such high energy use can only be justified in case of wastes that cannot be processed in another way (e.g. asbestos-containing). Therefore, the single-stage gasification process seems to be the better choice when dealing with small process streams (e.g. hazardous or medical waste). Additionally, the waste stream preferably consists of a large fraction of organic material, since the inorganic material is not necessarily converted into a slag.

A dual-stage gasification process can overcome some of the drawbacks related to the process combining gasification and plasma conversion in one reactor. Waste streams containing relatively large amounts of inorganic solid materials can benefit most from a dual-stage plasma gasification system. The high-temperature downstream plasma conversion step allows recuperation of valuable components, a significant reduction in volume and immobilization of the ash in a vitrified non-leaching slag that can be valorised as a by-product. The single-stage process has poor control of VOCs/tars, but the syngas cleaning step by the plasma in a dual-stage process assures an almost complete destruction of unwanted components in the gaseous end-products. Where a single-stage system is more suitable for smaller

89 C. Chapman, R. Taylor, and R. Ray, 2010; and M. Materazzi et al., 2016.

quantities of specific waste streams, dual-stage plasma gasification performs better for larger waste streams. The higher throughput relates to the fact that the decomposition of tars and chars in the syngas is much faster than the plasma decomposition of RDF.

The major drawback of plasma gasification, in general, is the high energy cost for generating the plasma. Electricity is an expensive energy vector, turning economic considerations into the strongest barrier for using plasmas for waste treatment. Not only the electricity cost in the OPEX of plasma torches is significant, but also the replacement cost for the electrodes is considerable. The short lifetime of the electrodes and the relatively low efficiency of the electric arc discharge plasma torches (about 50-70%) have led to an increasing interest in the use of microwave plasma torches in gasification. Lupa et al.[90] produced high-quality syngas from 10 g samples and proved the viability of microwave-induced plasma gasification. In a next step of that research, a demonstration facility, named Plasmergy, was erected by a collaboration between Stopford and Liverpool John Moores University. This installation processes up to 4 kg/h of C&IW through the combined action of three microwave plasma torches at nominal power levels of 6-8 kW. The conversion efficiency of electric energy to plasma enthalpy is relatively high at 80-90%. From the small-scale demonstration experiments, it is clear that microwave torches are advantageous over electric arc discharge torches in terms of energy efficiency and operational durability (and capital costs). Even though the microwave-induced plasma gasification has been demonstrated as a plausible WtE treatment method by Lupa et al.,[91] the viability in large-scale (commercial) gasification systems still needs to be proven.

3.3.4 *Energetic Valorization of Syngas*

Gasification is rarely installed as a stand-alone application, but is almost always followed by a downstream syngas valorization system. Besides the energetic valorization of syngas, discussed in this text, syngas can also be further converted to a range of fuels and/or chemicals (such as hydrogen, ammonia, methanol, etc.). The main reason to opt for waste gasification for energy purposes is that it can convert the waste to a secondary energy carrier that can be energetically valorized with a wider variety of highly efficient electricity-generating devices which can potentially increase the overall energy conversion efficiency of the process. To meet the specifications of high-efficiency, internally fired cycles (gas turbines, internal combustion engines) or fuel cells, the syngas needs to be cleaned from impurities. This type of gasification systems with a syngas-cleaning step before the power-producing step can be labelled as *power gasifiers*.[92] The other type of set-up is called *the heat gasifier*, where syngas is combusted in a boiler.

90 C.J. Lupa et al., 2013.
91 *Ibid.*
92 U. Arena, 2012.

3.3.4.1 Use in a Gas Boiler

This is the simplest pathway for energy recovery through gasification that does not require prior syngas cleaning and involves the direct burning of syngas in a boiler to feed an externally fired cycle or to be used as thermal energy directly. Except for the removal of coarse particulate matter, the combustion process poses no particular limitation on the level of undesired compounds. Any tars present in the syngas for example are completely oxidized. This type of gasification set-up is very similar to a combustion plant, with the difference that full oxidation is carried out in two steps: first feedstock gasification, then syngas combustion. Therefore, this "two-step oxidation" concept can be regarded as a particular type of a staged combustion plant, rather than a gasification plant.[93] Even if such processes do not allow capturing the full potential benefits of gasification, they are much simpler and easier to operate than a plant comprising syngas clean-up and downstream power production. Therefore, the heat gasifier configuration is adopted in most commercial technologies currently available. In addition to the previously mentioned advantages of gasification over combustion, the homogeneous gas phase combustion of syngas can be carried out under conditions more favourable than those achievable with raw waste.[94] However, the gasification plants tend to be more complex and costly, more difficult to operate and maintain, and less reliable. According to Consonni et al.,[95] the potential benefit that may justify the adoption of this type of gasification technology over conventional incineration plants is not the potential gain in electrical efficiency but is related to better combustion control with lower emissions and higher-quality recovery of metals and ashes (or slag). In the public opinion, this WtE gasification approach is often labelled as "indirect combustion" and such gasification plants as "incinerators in disguise", disfavouring public acceptance towards gasification.

3.3.4.2 Use in Internally Fired Cycles

In this category of power gasifier systems, the syngas must pass through a series of clean-up steps in order to remove species that would damage the downstream power generating device. A detailed discussion on syngas cleaning does not fall within the scope of this work, but a comprehensive review on the available cleaning technologies is made by Woolcock and Brown.[96] In order to utilize conventional low-temperature gas cleaning, such as acid gas removal, the syngas needs to be cooled down to approximately 100°C. It is important that tar has been removed before, otherwise condensation of sticky tar compounds occurs. The cooling can be accomplished by passing the syngas through a series of heat exchangers, which recover the sensible heat to produce steam (e.g. in a high-pressure steam boiler). This type of syngas cooling is called *boiler mode.*

93 S. Consonni and F. Viaganò, 2012.

94 U. Arena, 2012.

95 S. Consonni and F. Viaganò, 2012.

96 P.J. Woolcock and R.C. Brown, 2013.

The hot raw syngas can also be shock-cooled by water to saturation steam temperature (for example, by radial injection into a quench ring), called *quench mode*. The syngas is cooled almost instantaneously and exits with some of the quench water being vaporized in the process. The quenched syngas is saturated with water and must pass through a series of condensing heat exchanges that remove the moisture from the syngas (so that it can be recycled to the quench zone). Both gas engines and gas turbines, discussed here, have distinct advantages and disadvantages. Gas engines are known to be durable and reliable, but have a relatively low availability due to short maintenance intervals.

3.3.4.2.1 Gas Engine

The normal energy source in a gaseous-fuelled engine is methane. The quality of syngas as a fuel in internal combustion engines (ICE) is significantly poorer than gasoline and natural gas, especially because of the lower energy content. Additionally, a number of components of syngas (e.g. humidity and hydrogen levels) cause challenges such as corrosion and reduced efficiencies, and a predisposition to pre-ignition, respectively. Therefore, the engines require certain design modifications in order to be able to run on syngas. Gas engines running on syngas are mostly *spark-ignition* (SI) reciprocating engines. These engines require very little modification to run on syngas. The reduction in power of an engine run on syngas is mainly due to the lower net calorific value of the air/fuel mixture.[97] The power de-rates down to about 40-50%, with about 30% power loss due to the low energy density of syngas. The quality of gas should be sufficiently high in terms of low tar and particulates content to maintain the engine's consistent operation and to provide an adequate durability of major engine components. The allowed particle and tar concentration in producer gas for satisfactory operation of the internal combustion engine must be less than 50 and 100 mg/Nm^3, respectively.[98]

Compression ignition (CI) engines would operate more efficiently with syngas. However, syngas cannot be used in CI engines without a means of initiating combustion. Therefore, it is necessary to operate the engine in a dual-fuel mode, in which syngas is used as the primary fuel, and ignited by pilot diesel fuel. The de-rating of diesel engines running on producer gas is only between 15% and 30%.[99]

Gas engines can also be operated in combined heat and power mode by recuperating the sensible heat of the off gases in a heat exchanger to generate pressurized hot water or steam.

97 G. Sridhar, P.J. Paul, and H.S. Mukunda, 2001.
98 P. Hasler and T. Nussbaumer, 1999.
99 A. Pradhan, P. Baredar, and A. Kumar, 2011.

3.3.4.2.2 Gas Turbine

Gas turbines consist of a compressor unit, a combustion chamber and a turbine. Pressurized air coming from the compressor is mixed with the fuel stream (typically natural gas), after which the mixture is ignited, resulting in a high-temperature, high-pressure gas mixture. That mixture expands in a gas turbine, spinning a generator and producing electricity. Some key issues that need to be considered when using syngas rather than natural gas to fire a gas turbine are: flame stability issues, the level of water vapour, the lower calorific value and the higher corrosion potential.[100] The limitations on pollutants in the syngas are also more stringent than for gas engines. The particulate matter and tars cannot exceed 5 and 10 mg/Nm^3, respectively. Since syngas has a much lower LHV compared to natural gas, a higher mass and volume flow rate are required to generate the same power output. This has repercussions on the design of the gas turbine, namely the combustion chamber and the upstream equipment need to be adapted to these larger gas streams.

Syngas-fuelled gas turbines are almost exclusively used in a combined cycle (CC) power block (consisting of one or more gas turbines and a steam turbine). The high exhaust temperatures after the gas turbine, make it ideal for cogeneration. The excess heat from the gas turbine(s) is captured, converted into steam and sent to a steam turbine to produce additional electricity.

Moreover, the system is referred to as *integrated gasification combined cycle* (IGCC) when material and gas flows are internally linked between the gasification process and the heat- and electricity-generating unit. The process streams are interconnected resulting in an increased energy efficiency of the system.

3.3.4.2.3 Fuel Cell

Analogous to an IGCC power plant, *the integrated gasification fuel cell* (IGFC) cycle links a high-temperature-type fuel cell (such as the solid oxide fuel cell (SOFC)) power generation unit to the gasification process. Fuel cells offer the ability to convert chemical energy directly into electrical energy with a very low environmental impact.

The SOFC operates at higher temperatures than lower-temperature fuel cell types such as proton exchange membrane fuel cells or alkaline fuel cells. The latter fuel cell types require pure hydrogen as fuel, whereas the SOFCs are not poisoned by carbon monoxide and carbon dioxide, and can accept hydrogen, carbon monoxide, carbon dioxide, steam and methane mixtures as fuel directly, because of their internal shift and reforming capabilities.[101] The waste heat from SOFC systems may be captured and reused, increasing the theoretical overall efficiency to as high as 80-85%.[102] Recent devel-

100 T.B. Gibbons and I.G. Wright, 2009.
101 S. Giddey et al., 2012.
102 NASA, 2015.

opments allow an accessible incorporation of carbon capture and storage technology with IGCC and IGFC plants, which reduces the flow of CO_2 emitted to the atmosphere.[103]

The raw syngas contains tars and trace species such as particulates, chlorides (HCl), sulphur (HS) and alkali compounds among others, which could be detrimental to SOFCs if they are contained within the feeding syngas stream. Therefore, the syngas must be pre-treated in order to reduce these trace species to a level that SOFCs are able to withstand.[104] These fuel cell purity requirements are dependent on differences in the electrode material, microstructure and operating conditions, but generally demand a practically particulate-free syngas with levels of tar lower than 1 mg/m^3, sulphur-containing molecules lower than 1 ppm (even as low as 60 ppb), chloride lower than 100 ppmv and alkali compounds lower than a few ppm.[105]

3.4 Concluding Comparison of the Different WtE Technologies

The current chapter reviews thermochemical conversion technologies for the energetic valorisation of waste. The advantages and disadvantages of the four main types of thermochemical treatment technologies for MSW and/or RDF (i.e. incineration, pyrolysis, gasification and plasma gasification) are discussed in more detail in the following sections. In general, although some WtE processes accept raw (i.e. as received) MSW as input, the majority of WtE processes require a pre-processed form of MSW as the chemistry and fluid dynamics of more advanced technologies are very sensitive to feedstock variations (composition, humidity, ash content, particle size, density, reactivity, etc.). A more homogeneous waste input reduces variable and/or unstable operating conditions as well as fluctuating product qualities. Moreover, the more homogeneous properties allow for tighter process control, thus more stringent product quality specifications can be met.

Incineration processes provide a means to enable energy recovery of wastes without extensive pre-treatment. Municipal solid waste incinerators in particular offer a large potential source of heat/electricity, especially in the case where CHP is applied. The WtE possibilities for MSW incineration consist of recovering heat directly from the process and/or generating electricity using steam turbines. The flue gas quality does not comply with the high-quality standards required for gas turbines and gas engines, which produce electricity more efficiently than steam turbines. Nevertheless, state-of-the-art WtE incineration facilities can reach net electrical efficiencies of up to 26%.

---

103 N.S. Siefert and S. Litster, 2013.

104 M. Liu, 2013.

105 A. Bosmans, S. Wasan, and L. Helsen, 2013; W. De Jong and J.R. Van Ommen (eds.), 2015; and M. Sharma, N. Rakesh, and S. Dasappa, 2016.

Solid waste incinerators can obtain a significant waste volume reduction of about 90%[106] but still, a substantial volume of residues has to be disposed of mostly on landfills. Fly ash requires immobilization to make it environmentally safe for landfill disposal. Incinerator bottom ash is a highly heterogeneous mix of slag, ferrous and non-ferrous metals, ceramics, glass, other non-combustibles and residual organic matter. The potential risk of leaching of heavy metals has up till now restricted the reuse of bottom ash and it is still mostly disposed of in landfills. The conversion of MSW bottom ash into environmentally benign products calls for advanced thermal treatment options.

Pyrolysis produces a combustible gas, a solid product and – depending on the type of pyrolysis process – a liquid product. The gaseous stream consists primarily of CO and H_2, which makes it suitable for combustion in gas engines. Pyrolysis has demonstrated its feasibility for specific feed material (e.g. tires, electronic waste) but it does not offer a complete alternative to MSW incineration. The necessity of waste pre-treatment/after-treatment constitutes the main problem. Also, process conditions need to be optimized with respect to the specific waste input considered. The process viability is highly dependent on the presence of potential pollutants, which could make the pyrolysis products useless, or at least difficult to use, without additional treatment step(s). Due to the fact that (excavated) MSW is a highly heterogeneous material, the necessity to process MSW into a stream that complies with the stringent conditions inherent to pyrolysis processes is very likely to endanger the commercial viability of the process.

Gasification has several advantages over traditional combustion of MSW. It requires only a fraction of the stoichiometric amount of oxygen necessary for combustion. As a result, the formation of dioxins, SO_2 and NO_x is limited and the volume of process gas is low, requiring smaller and less expensive gas cleaning equipment. The lower gas volume also results in a higher partial pressure of contaminants in the off-gas, which favours more complete adsorption/absorption and particulate capture. Certain types of gasification reactors are capable of capturing the inorganic waste material in a vitrified non-leaching slag (e.g. the high temperature slagging gasifier). Moreover, waste volume reduction is increased compared to incineration. Finally, gasification generates a combustible gas that can be integrated with combined cycle turbines, gas engines and, potentially, with fuel cells for electricity (and heat) generation; these components produce electricity more efficiently than steam turbines. Unfortunately, syngas is still commonly burned in steam boilers. In Japan, less than 10% of the gasification plants generate electricity by high-efficiency systems after syngas cleaning.[107] Economical as well as practical considerations play a role here; as tar is burned in the combustor there is no need for extensive gas cleaning equipment which would add significantly to the plant investment costs. More advanced approaches, such as the emerging multi-stage gasification concepts, can possibly amend the tar issues and the ongoing developments are promising for the future.

106 C.R. Cheeseman et al., 2003.
107 G. Genon, V. Tedesco, and P. Urso, 2010.

Plasma gasification is a relative newcomer in the field of MSW/RDF treatment, but is becoming more and more attractive. The driving force is to give priority to environmental quality at affordable cost, hereby contributing to sustainable development. Plasma technologies use electricity as heat source rather than thermal energy liberated from combustion, which makes the heat source independent of the treated substances. This provides system flexibility, fast process control and more options in process chemistry, including the possibility of generating valuable products making it an attractive waste treatment option. Both single-stage and two-stage plasma gasification systems exist. In the single-stage design the waste is treated directly with plasma jets, while in the two-stage design, gasification is followed by plasma cleaning of the produced synthesis gas and/or residual solids fraction. Due to their high cost (electricity is an expensive energy vector), plasma systems have primarily been used for the treatment of several hazardous wastes (e.g. medical waste, low-level radioactive waste). In this field, plasma vitrification units have already demonstrated their technical feasibility.

Nowadays, plasma systems are becoming more accepted and their design is simplified, which means there is potential for widening its foothold in waste management as a viable alternative to more conventional treatment options. The cost savings associated with landfill avoidance and the added value of the potentially marketable products are the main economic advantages of plasma gasification and vitrification of wastes. From an environmental point of view, plasma systems have the potential to offer a viable alternative to landfilling and conventional incineration, with lower air pollution and virtually no residual waste streams requiring final disposal. The synthesis gas produced can be used for efficient electricity and/or heat generation, for conversion to second generation liquid (bio)fuels or for other applications. Residues from the synthesis gas cleaning processes (fly ash, precipitated metals, etc.) can be internally recycled and captured in the slag, which is vitrified and as such shows no leaching risks. This vitrified slag can be used as a construction material or even as a high-end secondary product (e.g. interlocking blocks, tiles and bricks). Within a regulatory context, combined plasma gasification and vitrification is regarded as a recovery process as opposed to a disposal technology, thereby increasing its social acceptance. The commercial viability of plasma technologies is expected to improve in the future, provided that regulatory, economic and sociopolitical drivers continue to encourage the implementation of advanced thermochemical conversion technologies.

References

Agon, N., 'Development and Study of Different Numerical Plasma Jet Models and Experimental Study of Plasma Gasification of Waste', 2015, Ghent University.

Apsimon, H., et al., 'Synergies in Addressing Air Quality and Climate Change', *Climate Policy*, Vol. 9, No. 6, 2009, pp. 669-80.

Aravind, P.V. and de Jong, W., 'Evaluation of High Temperature Gas Cleaning Options for Biomass Gasification Product Gas for Solid Oxide Fuel Cells', *Progress in Energy and Combustion Science*, Vol. 38, No. 6, 2012, pp. 737-64.

Arena, U., 'Process and Technological Aspects of Municipal Solid Waste Gasification. A Review', *Waste Management*, Vol. 32, 2012, pp. 625-39.

Bacq, A., et al., 'NOTAR – Novel Two Stage Downdraft Gasifier for Production of Clean Syngas for Industrial Applications', *International Conference on Renewable Energy Gas Technology*, Malmö (Sweden), 2016.

Bolhar-NordenKampf, M. and Isaksson, J., 'Refuse Derived Fuel Gasification Technologies for High Efficient Energy Production', *International Recycling and Recovery Congress*, Vienna (Austria), 2014.

Bosmans, A., Wasan, S., and Helsen, L., 'Waste to Clean Syngas: Avoiding Tar Problems', *Proceedings of the 2nd International Academic Symposium on Enhanced Landfill Mining*, Houthalen-Helchteren (Belgium), 2013.

Bridgwater, A.V., 'The Technical and Economic Feasibility of Biomass Gasification for Power Generation', *Fuel*, Vol. 74, No. 5, 1995, pp. 631-53.

Bridgwater, A.V., 'Renewable Fuels and Chemicals by Thermal Processing of Biomass', *Chemical Engineering Journal*, Vol. 91, No. 2-3, 2003, pp. 87-102.

Calabrò, P.S., 'The Effect of Separate Collection of Municipal Solid Waste on Lower Calorific Value of the Residual Waste', *Waste Management & Research*, Vol. 28, 2010, pp. 754-58.

Chapman, C., Taylor, R, and Ray, R., 'The Gas Plasma Process: Its Application in Enhanced Landfill Mining', *Enhanced Landfill Mining Symposium*, Houthalen-Helchteren (Belgium), 2010.

Cheeseman, C.R., et al., 'Ceramic Processing of Incinerator Bottom Ash', *Waste Management*, Vol. 23, No. 10, 2003, pp. 907-16.

Cheng, H. and Hu, Y., 'Municipal Solid Waste (MSW) as a Renewable Source of Energy: Current and Future Practices in China', *Bioresource Technology*, Vol. 101, 2010, pp. 3816-24.

Communication from the Commission to the European Parliament, the Council, the European Economic and Social Committee and the Committee of the Regions: Closing the loop – an EU action plan for the Circular Economy COM/2015/0614.

Consonni, S. and Viaganò, F., 'Waste Gasification vs Conventional Waste-to-Energy: A Comparative Evaluation of Two Commercial Technologies', *Waste management*, Vol. 32, 2012, pp. 653-66.

Council Directive 1999/31/EC of 26 April 1999 on the landfill of waste OJ L 182, 16.7.1999.

Damgaard, A., et al., 'Life-Cycle-Assessment of the Historical Development of Air Pollution Control and Energy Recovery in Waste Incineration', *Waste Management*, Vol. 30, 2010, pp. 1244-50.

De Jong, W. and Van Ommen, J.R. (eds.), *Biomass as a Sustainable Energy Source for the Future: Fundamentals of Conversion Processes*, Hoboken, NJ: John Wiley & Sons, Inc., 2015.

Directive 2000/76/EC of the European Parliament and of the Council of 4 December 2000 on the incineration of waste OJ L 332, 28.12.2000.

Directive 2008/98/EC of the European Parliament and of the Council of 19 November 2008 on waste and repealing certain Directives OJ L 312, 22.11.2008.

EIONET, 'European Topic Centre on Sustainable Consumption and Production', *What is Waste?*, 2012.

EMIS, 'Energie- & milieu-informatiesysteem voor het Vlaams Gewest', <www.emis.vito.be/> (accessed December 2016).

Energy Information Administration Office of Coal, Nuclear, Electric and Alternate Fuels – US Department of Energy, 'Methodoly for allocating Municipal Solid Waste to Biogenic and Non-Biogenic Energy', 2007.

Energy Research Centre of the Netherlands (ECN), 'Phyllis – Database on Composition of Biomass and Waste', 2011.

European Commission, 'Reference Document on Best Available Techniques for the Waste Treatment Industries', 2006.

Eurostat, 2014, available at <http://appsso.eurostat.ec.europa.eu/nui/submitViewTableAction.do>.

Eurostat, 2016, available at <http://ec.europa.eu/eurostat/statistics-explained/index.php/File:Gross_electricity_production_by_fuel,_GWh,_EU-28,_1990-2015-T1.png>.

Fauchais, P. and Vardelle, A., 'Thermal Plasmas', *IEEE Transactions on Plasma Science*, Vol. 25, 1997, pp. 1258-80.

Feiner, J., Mi, S., and Schmid, W., 'Meeting the Challenges of Future Urbanization: Risks and Opportunities for Future Urban Planning in the People's Republic of China', *Disp*., Vol. 145, 2001, pp. 10-18.

Folkelid, M.N., 'Cortus Energy: WoodRoll – Ultraclean Syngas Taking Bioenergy Forward', *Third Conference of the European Biogas Association*, Ghent (Belgium), 2016.

Genon, G., Tedesco, V., and Urso, P., 'Assessment of the Feasibility of an Innovative Technology Plant Aimed at the Energetic Valorization of Municipal Solid Waste in the Province of Turin', *Third International Symposium on Energy from Biomass and Waste*, Venice (Italy), 2010.

Gibbons, T.B. and Wright, I.G. – Oak Ridge National Laboratory, 'A Review of Materials for Gas Turbines Firing Syngas Fuels', 2009.

Giddey, S., et al., 'A Comprehensive Review of Direct Carbon Fuel Cell Technology', *Progress in Energy and Combustion Science*, Vol. 38, No. 3, 2012, pp. 360-99.

Gohlke, O., 'Efficiency of Energy Recovery from Municipal Solid Waste and the Resultant Effect on the Greenhouse Gas Balance', *Waste Management & Research*, Vol. 27, 2009, pp. 894-906.

Gomez, E., et al., 'Thermal Plasma Technology for the Treatment of Wastes: a Critical Review', *Journal of Hazardous Materials*, Vol. 161, No. 2-3, 2009, pp. 614-26.

Grant, M., 'Enerkem Biorefineries: Setting a New Global Standard in Biofuels, Chemicals and Waste Management', *Gasification and Syngas Technology Council*, Vancouver (Canada), 2016.

GreenE, 'GreenE – Waste to Energy Gasification System', <http://greene.es/> (accessed 21 October 2016).

Guddeti, R.R., Knight, R., and Grossmann, E.D., 'Depolymerization of Polyethylene Using Induction-Coupled Plasma Technology', *Plasma Chemistry and Plasma Processing*, Vol. 20, No. 1, 2000, pp. 37-64.

Gupta, A.K. and Cichonski, W., 'Ultrahigh Temperature Steam Gasification of Biomass and Solid Wastes', *Environmental Engineering Science*, Vol. 24, 2007, pp. 1179-89.

Hasler, P. and Nussbaumer, T., 'Gas Cleaning for IC Engine Applications from Fixed Bed Biomass Gasification', *Biomass and Bioenergy*, Vol. 16, No. 6, 1999, pp. 385-95.

Heberlein, J. and Murphy, A.B., 'Thermal Plasma Waste Treatment', *Journal of Physics D: Applied Physics*, Vol. 41, No. 5, 2008, pp. 053001 (20pp).

Heidenreich, S. and Foscolo, P.U., 'New Concepts in Biomass Gasification', *Progress in Energy and Combustion Science*, Vol. 46, 2015, pp. 72-95.

Helsen, L., 'Low-Temperature Pyrolysis of CCA Treated Wood Waste', 2000, KU Leuven.

Higman, C., 'State of the Gasification Industry – The Updated Worldwide Gasification Database', *International Pittsburgh Coal Conference*, Beijing (China), 2013.

Hoel, M. and Kverndokk, S., 'Depletion of Fossil Fuels and the Impacts of Global Warning', *Resource and Energy Economics*, Vol. 18, No. 2, 1996, pp. 115-36.

Institute for Energy Research, 'China to Build 50 Coal Gasification Facilities', 2014.

Jones, P.T., et al., 'Enhanced Landfill Mining in View of Multiple Resource Recovery: A Critical Review', *Journal of Cleaner Production*, Vol. 55, 2013, pp. 45-55.

Juniper Consultancy Services Limited, 'Alter NRG/Westinghouse Plasma Gasification Review', 2008.

Kar, B., 'Clean Fuels from Coal/Petcoke Using Shell Gasification Technology', *Conference on Clean Fuels from Coal/Petcoke*, New Delhi (India), 2006.

Karak, T., Bhagat, R., and Bhattacharyya, P., 'Municipal Solid Waste Generation, Composition and Management: the World Scenario', *Critical Reviews in Environmental Science and Technology*, Vol. 42, No. 15, 2012, pp. 1509-630.

Klein, A., 'Gasification: an Alternative Process for Energy Recovery and Disposal of Municipal Solid Wastes', 2002, Columbia University.

Kleis, H. and Dalagar, S., *100 Years Of Waste Incineration in Denmark – from Refuse Destruction Plants to High-technology Energy Works*, Denmark: Babcock & Wilcox Vølund, 2007.

Kolb, T. and Seifert, H., 'Thermal Waste Treatment. State of the Art – A Summary', *Waste Management 2002: The Future of Waste Management in Europe*, Strasbourg (France), 2002.

Lemmens, B., et al., 'Assessment of Plasma Gasification of High Calorific Waste Streams', *Waste Management*, Vol. 27, No. 11, 2007, pp. 1562-69.

Limerick, 'Feasibility Study of Thermal Treatment Options for Waste in the Limerick/Clare/Kerry Region', 2005.

Liu, M., 'Biomass-Powered Solid Oxide Fuel Cells', 2013, Delft University of Technology.

Lombardi, L., Carnevale, E., and Corti, A., 'A Review of Technologies and Performances of Thermal Treatment Systems for Energy Recovery from Waste', *Waste Management*, Vol. 37, 2015, pp. 26-44.

LTD., EBARA Environmental Plant CO., 'Pressurized Twin Internally Circulating Fluidized-Bed Gasification System (PTIFG)', <www.eep.ebara.com/en/products/gas.html> (accessed 23 November 2016).

Lupa, C.J., et al., 'The Use of Commercial and Industrial Waste in Energy Recovery Systems – a UK Preliminary Study', *Waste Management*, Vol. 31, No. 8, 2011, pp. 1759-64.

Lupa, C.J., et al., 'Gas Evolution and Syngas Heating Value from Advanced Thermal Treatment of Waste Using Microwave-Induced Plasma', *Renewable Energy*, Vol. 50, 2013, pp. 1065-72.

Machiels, L., et al., 'Properties of Inorganic Polymer Cement from Ferric and Ferrous Vitrified Residues of Plasma Gasification', *Slag Valorisation Symposium – Zero Waste*, Leuven (Belgium), 2015.

Malkow, T., 'Novel and Innovative Pyrolysis and Gasification Technologies for Energy Efficient and Environmentally Sound MSW Disposal', *Waste Management*, Vol. 24, No. 1, 2004, pp. 53-79.

Materazzi, M., et al., 'Performance Analysis of RDF Gasification in a Two-Stage Fluidized Bed-Plasma Process', *Waste Management*, Vol. 47, 2016, pp. 256-66.

Min, B.Y., et al., 'Study on the Vitrification of Mixed Radioactive Waste by Plasma Arc Melting', *Journal of Industrial and Engineering Chemistry*, Vol. 13, No. 1, 2007, pp. 57-64.

Mondal, P., Dang, G.S., and Garg, M.O., 'Syngas Production through Gasification and Cleanup for Downstream Applications – Recent Developments', *Fuel Processing Technology*, Vol. 92, 2011, pp. 1395-410.

NASA, 'Solid Oxide Fuel Cells & Electrolysis Membranes', <www.grc.nasa.gov/WWW/StructuresMaterials/Ceramics/research_solid.html> (accessed 15 April 2015).

National Energy Technology Laboratory – US Department of Energy, 'Advantages of Gasification – Feedstock Flexibility', 2012.

Nema, S.K. and Ganeshprasad, K.S., 'Plasma Pyrolysis of Medical Waste', *Current Science*, Vol. 83, No. 3, 2002, pp. 271-78.

Niccolucci, V., et al., 'Biocapacity vs Ecological Footprint of World Regions: a Geopolitical Interpretation', *Ecological Indicators*, Vol. 16, No. SI, 2012, pp. 23-30.

Office of Solid Waste – United States Environmental Protection Agency, 'Municipal Solid Waste in the United States: 2011 Facts and Figures', 2013.

Pradhan, A., Baredar, P., and Kumar, A., 'Syngas as an Alternative Fuel Used in Internal Combustion Engines: A Review', *Journal of Pure and Applied Science & Technology*, Vol. 5, No. 2, 2011, pp. 51-66.

Quaghebeur, M., et al., 'Characterization of Landfilled Materials: Screening of the Enhanced Landfill Mining Potential', *Journal of Cleaner Production*, Vol. 55, 2013, pp. 72-83.

Reimann, D.O. – Confederation of European Waste-to-Energy Plants (CEWEP), 'Energy Report II (Status 2004-2007)', 2009.

Reimann, D.O. – Confederation of European Waste-to-Energy Plants (CEWEP), 'Energy Report III Status 2007-2010', 2012.

Ruiz, J.A., et al., 'Biomass Gasification for Electricity Generation: Review of Current Technology Barriers', *Renewable and Sustainable Energy Reviews*, Vol. 18, 2013, pp. 174-83.

Sandeep, K. and Dasappa, S., 'Oxy-Steam Gasification of Biomass for Hydrogen Rich syngas Production Using Downdraft Reactor Configuration', *International Journal of Energy Research*, Vol. 32, No. 2, 2014, pp. 174-88.

Schabbach, L.M., et al., 'Valorization of MSWI Bottom Ash through Ceramic Glazing Process: A New Technology', *Journal of Cleaner Production*, Vol. 23, No. 1, 2012, pp. 147-57.

Sharma, M., Rakesh, N., and Dasappa, S., 'Solid Oxide Fuel Cell Operating with Biomass Derived Producer Gas: Status and Challenges', *Renewable and Sustainable Energy Reviews*, Vol. 60, 2016, pp. 450-63.

Siefert, N.S. and Litster, S., 'Exergy and Economic Analyses of Advanced IGCC-CCS and IGFC-CCS Power Plants', *Applied Energy*, Vol. 107, 2013, pp. 315-28.

Sridhar, G., Paul, P.J., and Mukunda, H.S., 'Biomass Derived Producer Gas as a Reciprocating Engine Fuel – An Experimental Study', *Biomass and Bioenergy*, Vol. 21, No. 1, 2001, pp. 61-72.

Steiner, C., et al., 'EBARA's Fluidized Bed Gasification: Atmospheric 2x225 t/d for Shredding Residues Recycling and Two-Stage Pressurized 30 t/d for Ammonia Synthesis from Waste Plastics', *2nd International Symposium on Feedstock Recycling of Plastics*, Oostende (Belgium), 2002.

Tanigaki, N., 'Proven Waste Gasification Technology for Circular Society', *World Waste to Energy City Summit*, London (UK), 2015.

Taylor, R, Chapman, C., and Faraz, A., 'Transformations of Syngas Derived from Landfilled Wastes Using the Gasplasma Process', *2nd International Symposium on Enhanced Landfill Mining*, Houthalen-Helchteren (Belgium), 2013.

Turns, S.R., *An Introduction to Combustion: Concepts and Applications*, Boston: McGraw-Hill, 2000.

UBA, 'Draft of a German Report with Basic Informations for a BREF-Document. Waste Incineration', 2001.

United Nations – Department of Economic and Social Affairs Population Division, 'World Urbanization Prospects: The 2014 Revision, Highlights', 2014.

Valero, A. and Uson, S., 'Oxy-co-gasification of Coal and Biomass in an Integrated Gasification Combined Cycle (IGCC) Power Plant', *Energy*, Vol. 31, 2006, pp. 1643-55.

Van den Loosdrecht, J. and Niemantsverdriet, J.W., 'Synthesis Gas to Hydrogen, Methanol and Synthetic Fuels', in R. Schloegl (Ed.), *Chemical Energy Storage*, Berlin: De Gruyter, 2013.

Wagland, S.T., et al., 'Comparison of Coal/Solid Recovered Fuel (SRF) with Coal/Refuse Derived Fuel (RDF) in a Fluidised Bed Reactor', *Waste Management*, Vol. 31, No. 6, 2011, pp. 1176-83.

Warnecke, R., 'Gasification of Biomass: Comparison of Fixed Bed and Fluidized Bed Gasifier', *Biomass and Bioenergy*, Vol. 18, No. 6, 2000, pp. 489-97.

Westinghouse Plasma Corporation, 'Westinghouse Plasma Gasification is the Next Generation of Energy-from-Waste Technology', *Energy from Waste*, 2012.

Woolcock, P.J. and Brown, R.C., 'A Review of Cleaning Technologies for Biomass-Derived Syngas', *Biomass and Bioenergy*, Vol. 52, 2013, pp. 54-84.

Working Group Energy Recovery (Ramboll) – ISWA, 'Waste to Energy: State of the Art Report, 6th Edition', 2012.

Zevenhoven, R. and Saeed, L. – Helsinki University of Technology Department of Mechanical Engineering, 'Automotive Shredder Residue (ASR) and Compact Disc (CD) Waste: Options for Recovery of Materials and Energy', *Energy Engineering and Environmental Protection Publications*, 2003.

4 WtE in Practice: The Clean and Safe Sink Solution of the Amsterdam Waste Energy Company (AEB)

Sietse A. Agema and Matthias Wieseler

4.1 Introduction

> Mass burning of waste to reduce volume of waste which otherwise needs to go to landfill in order to secure public health by destroying all toxic pathogenic organic particles and creating energy.

Processing residual waste streams for society means solving problems caused by societies, where the production and the use of products have the function of enhancing the economy and the societies' welfare in general. In this paradigm, solutions should be found that are robust and that come at the lowest societal cost possible, causing no environmental problems now, nor in the future.

When we really start considering the impact on the earth's resources, a more dedicated and circular approach is necessary. This will require an effort by the whole of society in designing, producing, consuming, returning and processing these functions/commodities. It is an overall gain to eliminate inefficiencies in our overall production and consumption scheme. Since this is not an overnight exercise, a robust sink solution will be necessary for this transition, but probably also after that.

4.2 Background to Waste to Energy

The concept of waste-to-energy (WtE) as a sustainable alternative to landfilling household or residual waste is increasingly gaining acceptance. In ecologically advanced parts of Europe such as the Netherlands, Switzerland or Scandinavian countries, the most modern waste management systems in the form of grate-based WtE plants have been installed. These plants achieve high recovery rates and produce electricity, process steam and district heating.[1]

1 Berlo, *et al.*, 2011.

Waste is mainly dealt with as a problem of hygiene and other health-related risks. Only recently, with the development of the circular economy model, the potential of waste as a resource is taken as the starting point for waste management regulations.[2] WtE tended to be principally designed around the paradigm "design to be clean", designed to minimize the quantity and the (negative) effect of its emissions. There is, however, a new and strong tendency to develop a new generation of WtE that are "designed for output", maximizing the recovery of energy and materials. This requires new evaluations of the effects of the outputs of WtE.[3]

The residual municipal solid waste (MSW) and similar commercial waste, are difficult streams that remain after all other possibilities of the waste hierarchy are exhausted. For this waste, the normal choices are landfilling or combustion (incineration) in WtE plants. Of course, incineration should only be considered as the alternative for landfilling. Only an effective outlet for the difficult material will allow breaking the impossible competition of reuse, recovery and recycling (RRR) with low landfilling prices. This is so, because of the relative high price, WtE is never a real competitor to RRR. But by avoiding the cheap outlet to landfilling, it is an effective stimulus for RRR. This is also shown by the data of countries in the EU where a high RRR corresponds perfectly to the installed WtE capacities.

Figure 1 EU27 recycling, burning and landfill

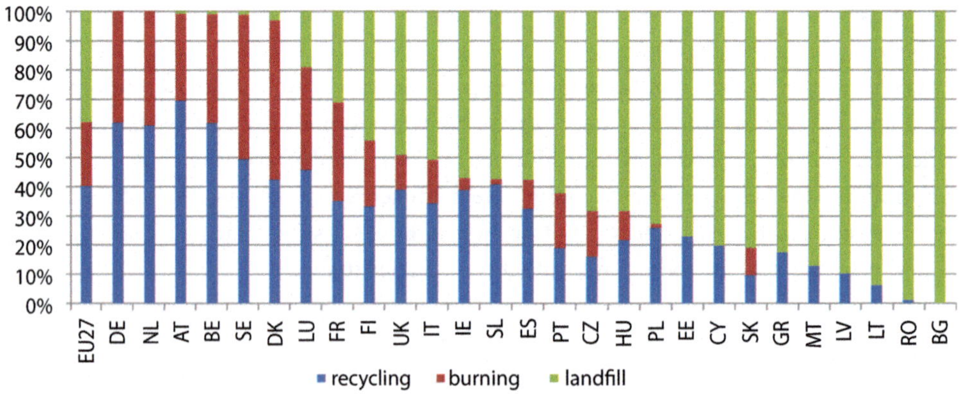

2 Ellen Mcarty Foundation.
3 Berlo, *et al.*, 2008.

Figure 2 CO_2 evaluation

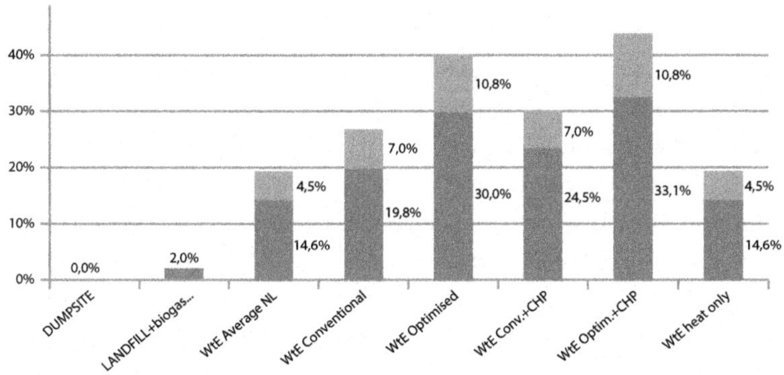

A CO_2-evaluation is made for landfill and WtE concepts.[4] Different concepts are identified and compared for their performance on energy and materials recovery. Performance indicators for WtE are compared; like energy efficiency, exergy efficiency, the R1-D10 formula from the EU Waste Framework directive, and CO_2-emission and avoidance.

It is shown that, due to the biomass content and the avoidance effect caused by the recovery of energy and materials, conventional WtE has a near-zero CO_2 emission per tonne of waste. Optimized WtE can have a significant negative overall emission of 200-300 kg CO_2/tonne of waste. This means an absolute net avoidance of CO_2 by WtE. The reduction relative to landfilling is as much as 500-1,200 kg CO_2/tonne of waste. The potential for optimization of the energy recovery as well as the material recovery of the WtE infrastructure is demonstrated. If WtE is evaluated as a power plant, an optimized plant can have an emission of only 0.336 kg CO_2/kWh, lower than a gas-fired electrical power plant, and this absolute figure does not include the avoided landfill emissions. With combined heat and power (CHP), this can be reduced even further.

The actual potential of electricity production from WtE for the EU15 is calculated to be over 7.5% of total electricity production. Additionally, heat and the metal recoveries could be doubled.[5]

4.3 A Short History

Amsterdam lies in a part of Holland with a long tradition of reclaiming land from the sea and lakes. A complex water management system was created, in order to keep people's feet dry. In the early days, the canal system was used for discharging sewage and residual waste streams. Already in the fifteenth century, the first waste bins are mentioned as the city council tried to keep the city clean and tidy; smell was the main driver in those days.

4 *Id.*
5 *Id.*

This smell caused many complaints by citizens. The waste composition was mainly determined by kitchen waste, dung and ash which at that time was often collected separately and sometimes used as animal feed or fertilizer for the agricultural sector. The contracts for street cleaning and the disposal of waste were left to the private sector. Fines were imposed on people who tried to use the canals as a public toilet. In 1831-1832, the first cholera epidemic arrived from Asia and hit the city with 1,200 casualties. It struck again in 1848-1849. Now doctors became involved in the discussion about waste, since they were able to identify a link between hygiene and the epidemic. So a lesson that Amsterdam learned from that period is that waste management and public health have a close relationship with each other. In 1877, Amsterdam realized that a new institution responsible for taking care of waste produced by its citizens was required. This could no longer be left to private parties since the income from ashes, organic waste and dung was no longer sufficient. Residual waste streams were collected and brought outside of the city and dumped into dried lakes, known as polders. But it was soon recognized that this practice was not sustainable due to space limitations and hazardous potential caused by the contamination of the waste and groundwater, since protection layers were not used at that time. Already in 1852 the first design of a waste incinerator was made, a project that never came to financial close.

4.3.1 First Generation WtE Plant

Later on in 1917, Amsterdam decided to build a WtE facility, primarily to reduce the amount of waste produced by its citizens by volume and mass. It followed good examples of London and Hamburg. By burning waste, the toxic organic fractions are reduced. But the capacity was not enough to treat all the waste produced and some landfilling remained in practice. But the incineration of waste also caused problems. During the interbellum and post-war development, the composition of waste had changed and had become even more hazardous. Components had been introduced that produce other hazardous exhaust gases through the burning process. The first generation of waste incinerators consisted of basic burning units with simple energy recovery in the form of electricity. Hardly any consideration for flue gas cleaning was in place, resulting in black smoke from chimneys polluting the environment as was common in the early times of the industrial revolution.

4.3.2 Second Generation WtE Plant

Then, in 1963, a modern 2e generation facility was built next to the old facility as a first step for flue gas cleaning. It was introduced by means of an electrostatic precipitator

(ESP) in order to capture some of the dust. This facility caused a big scandal in the 1980s. As we now know, it produced not only dust. Dioxins discovered in milk produced in the surroundings of Amsterdam were found to be directly linked to the waste incinerator.

4.3.3 Third Generation WtE Plant

Due to this 1980s scandal, the third generation WtE facility was designed to be clean. No discussion of the environmental performance should occur for this newly designed WtE facility. The newly developed emission limits from the permitting authority, pushed the designers to do their utmost in order to design a cleaning facility for the flue gasses. With success, as they have now reached levels that are even lower than allowed. When this emission and dioxin problem was solved, WtE ceased to be the subject of societal discussion and came to be regarded as an accurate sink solution. The third generation was owned by 19 municipalities, in order to create enough volume to reach a sufficient economy of scale. In that period, it was expected that the amount of MSW would still increase. At the time, municipalities produced roughly 600,000 tonnes of MSW per year, expecting the volume to increase to 800,000 t/y. Fortunately, recycling was becoming popular and regulations made it obligatory to reduce the amount of waste down to 450,000 tonnes. The additional waste streams burned in this facility are called Commercial Waste. It is collected at industries and offices and includes more and more residues from recycling processes. So, especially in urban areas where carrying out separated collection with the technologies of that time was hard to achieve, WtE became accepted as a sink solution. Now the environmental issues were solved. Proof came in 2015 after an extensive biomonitoring programme executed by the Wageningen University (WUR). The results showed that the emissions did not affect the quality of crops and cow milk. Concentrations of heavy metals, PAHs and dioxins/PCBs were generally similar to background levels and did not exceed standards for maximum allowable concentrations in foodstuffs (e.g. vegetables and cow milk).[6] This programme was executed around three WtE facilities in the Netherlands. AEB Amsterdam was not among them due to the fact that too many other emission-producing industries are in the vicinity of the plant, which could cause problems in determining the source of the producer. But it is known that AEB Amsterdam has a better performance on these emissions than the three tested, so the same result or better is to be expected.

6 Van Dijk, *et al.*, 2015.

4.3.4 Fourth Generation WtE Plant

In 2003, AEB Amsterdam started to design the fourth generation of WtE, intended for maximum recovery. It was an answer to the moratorium set by the Dutch government on mass burning with grate ovens. The moratorium was introduced because at the time, gasification and the use of fluidized beds was thought of as the best new technology for waste incineration.

The high-efficiency technology developed by AEB promised 30% net electrical efficiency, while grate incineration at that time could not achieve more than 23% net electrical efficiency. Since then, it has been proven that high-efficiency technology and grate ovens can easily achieve more than 31% net electricity effect, even without any pre-treatment. The high-efficiency technology proved also to have an availability of greater than 92% over the last 9 years. Before this period, fluidized bed still needed extensive pre-treatment (so no sink) and even then they did not reach an availability of 90%. This is still the case today. Meanwhile, gasification is also still not proven to be a robust technology for untreated MSW, so it cannot be considered a sink technology.

Table 1 Waste to Energy facilities as power plants – changing paradigms for WtE-plants

Generation	Period	Paradigm	Boiler	Heat efficiency	Electrical efficiency
0	Pre-1900	*Landfill*	-	-	-
1st	1900-1960	*Sanitation*, Incineration	Hot water or low steam parameters	~50%	Some cases
2nd	1960-1990	*Amenities*, Dust separation	Medium steam parameters	<70%	~5-14%
3rd	1990-2005	*Pollution*, Chemical Flue gas cleaning	40 bar/400C	CHP	~15-22%
4th	2005-	*Sustainability*, Energy efficiency	Higher steam parameters	CHP	~25-31%

4.3.5 Transforming AEB into a Commodity Company

The next step is for Amsterdam to work on improving its separation rates, which at this moment in time are at 27% of the total of MSW. The aim is to reach 65% in 2020. This will partly be done by increasing source separated streams by intensifying the collection infrastructure for textiles, paper, glass, plastics/drinking cartons and organic waste. But the highest rates are expected in the post-separation plant, which is to build on the AEB premises in order to reach a separation rate of at least 50% (the plant will start operating at the end of 2017). Other initiatives focus on other treatments of these streams in order to maintain their functionality in society as long as possible.

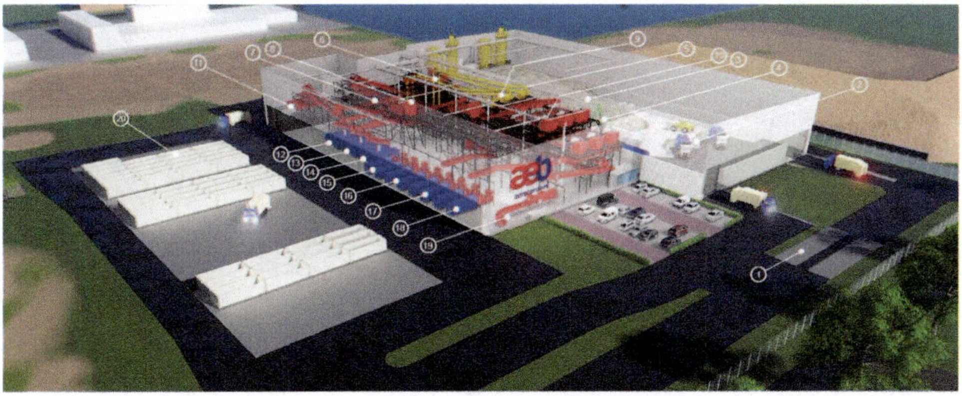

Another development is on the inert fractions coming from the burning process. First, all ferro will be filtered out of the inert fraction. Now the inert fraction due to its contamination – and stringent Dutch regulations – is only to be used for simple road construction. As it is now, it is used in dikes protected by layers in order to prevent leaking. At the moment, AEB Amsterdam is building a large-scale demo plant to continue the successful pilots where this inert fraction is transferred into clean commodities like granulate, sand and even cement (the production of cement is still in its pilot phase) In this transition, CO_2 is in a sophisticated "aging" process where heavy metals are immobilized and CO_2 is fixed in the product. The process is cost effective compared to its current application.

4.4 Boundaries in the Circular Economy

In a circular economy, the aim is to recover materials and energy on the highest level possible. For products, this means to reuse them. In a circular economy, it is further the aim to eventually produce in such a way that a particular function will last and can be repaired easily, or materials can be reused at a very high quality level so that residues will be limited to the smallest amounts possible.

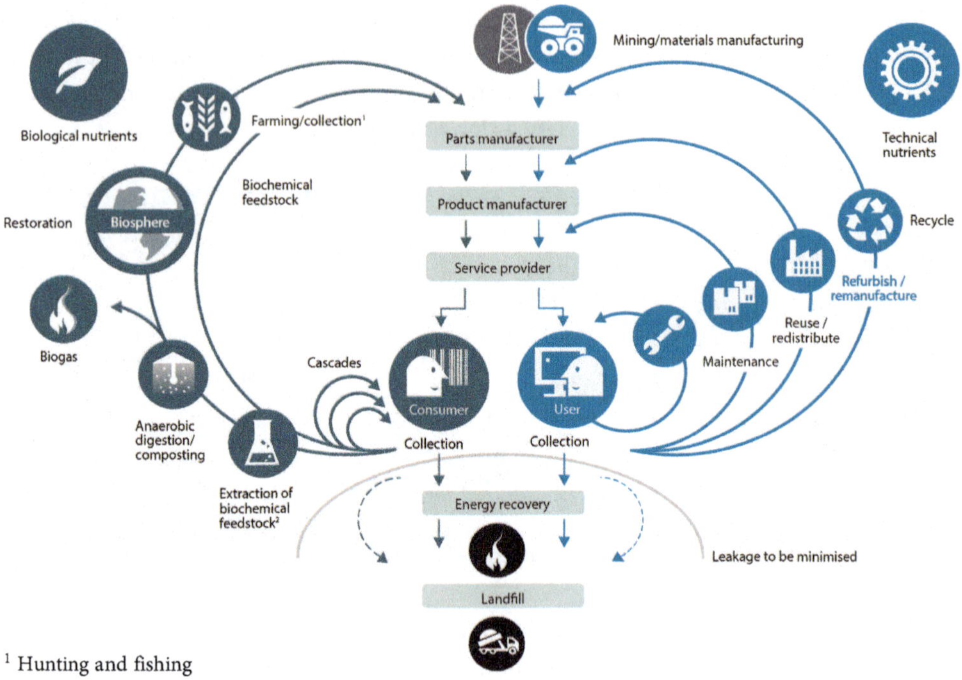

[1] Hunting and fishing

[2] Can take both post-harvest and post-consumer waste as input

Source: Ellen MacArthur Foundation circular economy team

This is good news for waste management. The greater reduction of virgin materials and the reuse function leaves less residual waste streams for energy recovery. But how long this transition will take is hard to estimate. The main economic drivers for a circular economy are scarce commodities. Some commodities, like oil, will become scarce in a few hundred years (EIA, 2010) while 90% of this commodity is currently used one way as fuel for transport and energy. At the moment, there is a global growth in the chemical industry and in the manufacturing of plastic goods, with decreasing prices worldwide.

We also see that scarcity is not directly an economic driver. Phosphate mines are expected to be exhausted within 20 years, yet there is no effect on the commodity market so far, even though everyone is aware of the essential uses of phosphates. It means that even the relatively short 20-year horizon is not short enough to become a powerful economic driver.

Another issue is the reuse potential of waste streams. Even if the quality of the recycled waste stream is high, the recycle rate (the times that a polymer fibre or other material can be reused without loss of quality) is limited for some materials. A good example for the limits of recycling materials is found in the "used paper" case. The reuse of source separate collected paper still makes it necessary to add 20% of new cellulose-fibres in the paper process, in order to maintain quality. It means that 20% is often used for thermal recovery of which the energy is used for partly maintaining its own process.

Furthermore, some polymers are expected to be reused five times before their function gets lost. Therefore, when it comes to material recovery instead of reuse of function, we will see the degradation of materials eventually following the law of physics into energy. And when this reverts to energy through incineration, we should make sure that we reclaim the highest amount of energy possible, aiming for maximum efficiency at the lowest societal costs. So the real successful change for the circular economy must come from the way we are to consume differently, changing from the possession of goods to using goods as a service. So a paradigm shift for producers is to change from design-to-fail to design-to-last, where the economic value is in the lease of products instead of selling them directly.[7]

4.5 Potential High-Efficiency Technology as Sink Solution in a Circular Economy

Waste to energy can never be in full competition with recycling. It should focus on treating materials that cannot be cost effectively recycled due to their composition or way of use. One problem is that some materials get highly contaminated with other polluting sources through the way they are being used or collected. One example is paper when it is collected in mixed residual household waste, or plastics with multiple layers of different materials. The main reason for the contamination of materials is the collection method for residual waste streams in greater urban areas, where separate collection at source is hard to implement due to a lack of public and private space. Meanwhile, post-separation plants currently reach a 50% recycling rate at the most.

In a recent study, the two major sink solutions for waste management were compared and evaluated.[8] It is shown that improved energy efficiency as in the high-efficiency technology has the highest impact on CO_2 reduction. The reason for this is that energy generated from waste replaces electricity produced by an energy mix (energy mix of a country is the average of energy produced by: wind, solar, coal, gas and other sources specific to a country). Taking into account that MSW contains 54% materials with bio-mass origin it is a sustainable renewable energy source. The replacement of electricity production from an energy mix with electricity produced with a 54% renewable source means a huge CO_2 reduction. Besides energy recovery, material recovery as in metals and non-metals from the slacks (burning residue), or inert fraction of the MSW, also adds to the reduction of CO_2 emissions compared to landfilling.

7 Rau, *et al.*, 2016.
8 Berlo, *et al.*, 2008.

Figure 3 Greenhouse effect overall

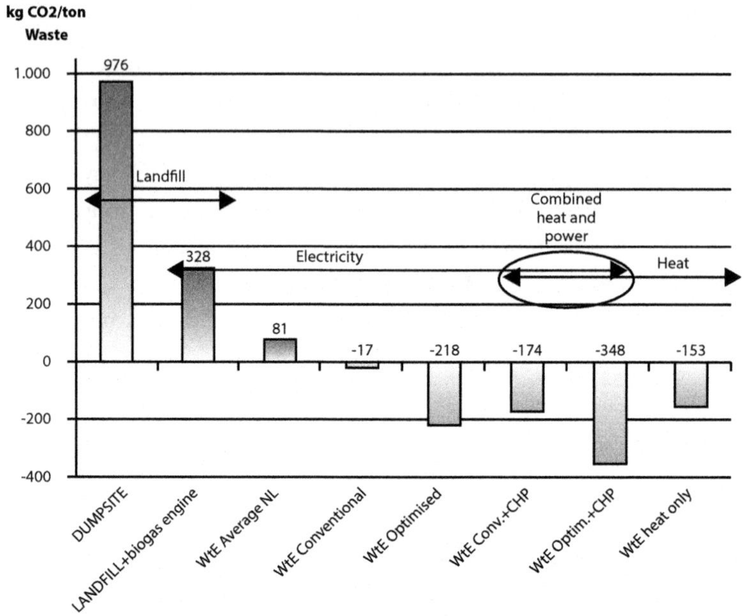

Therefore, WtE high-efficiency technology adds up to recycling if the residual outputs of the recycling industry will be transferred into energy with the highest energy gain, because the life cycle analysis of a recycling method is heavily determined by the way the residuals are treated.

If we assess European figures, we see a huge potential for increasing the use of WtE technology as long as we are able to divert waste from landfill. If we use a high efficiency process >30% in order to convert these waste streams into energy, the potential for the reduction in CO_2 emissions is staggering and could reach 200 million tonnes of CO_2 per year. This is only the electrical part, which covers 7.5% of the total electricity consumption in Europe.[9]

4.6 High-Efficiency Technology Explained

The Amsterdam plant with its two new lines in the High Efficiency (HRC) block was designed for particularly high energy recovery with respect to electricity generation. The challenge of achieving this goal was considerable, since to date, boilers of 40 bar at 400°C were deemed to be the most suitable for WtE given the constraints posed by corrosion in the furnace and boiler. Increasing live steam temperature creates superheater corrosion and increasing boiler pressure with a given live steam temperature is limited by acceptable

9 *Id.*

steam wetness in the low-pressure turbine. The new HRC lines with a total waste treatment capacity of 530.000 t/year @10 MJ/kg were commissioned in 2007 using a very efficient energy concept consisting of a 130-bar boiler operating with superheated steam at 440°C and intermediate reheater.[10] The steam–steam–reheater allows a strong increase in steam pressure with moderate live steam temperature. This enables a new range of steam parameters and a new flexibility in shifting boiler load from superheaters to radiation part. This allows for gross electrical efficiency of 34.5% and a net design efficiency of just over 30%.[11]

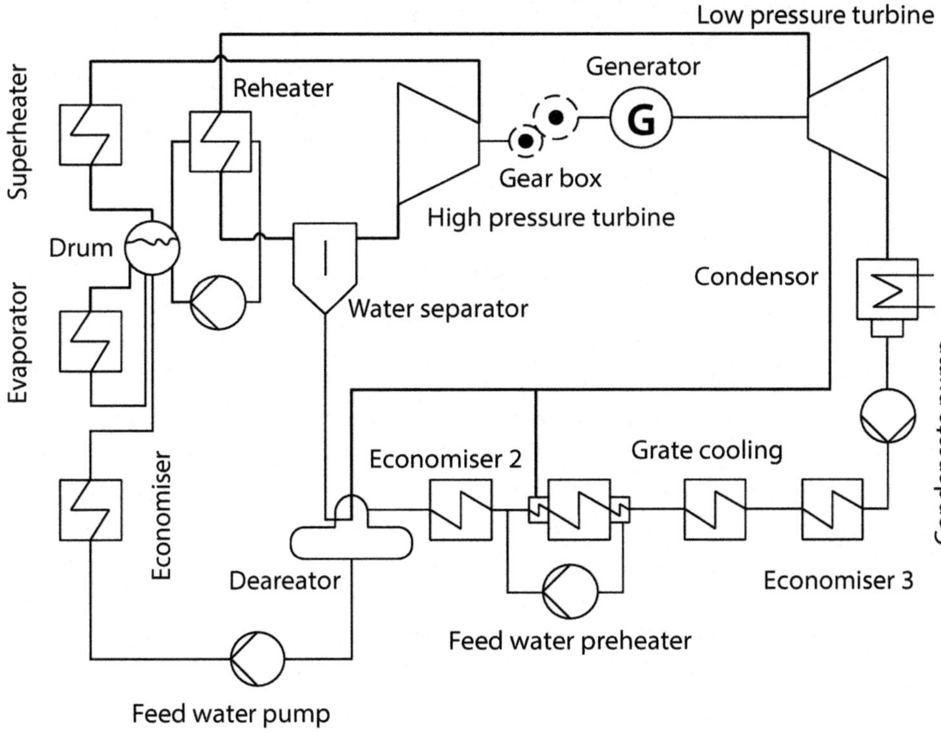

The AEB high-efficiency technology relies on a high net–gross efficiency combined with a high availability. In the design phase, the main goal parameters were stated as

- Overall net–gross efficiency >30%
- Overall availability >90%

The efficiency is primarily achieved through a reheater cycle in the steam turbine and substantially higher steam parameters, compared to conventional plants. The required

10 AEB 2007.
11 Berlo, *et al.*, 2011.

availability can be reached by avoiding corrosion in the boilers and the measures taken in the concept are design rules for superheaters, the flue gas path and air injection parameters.

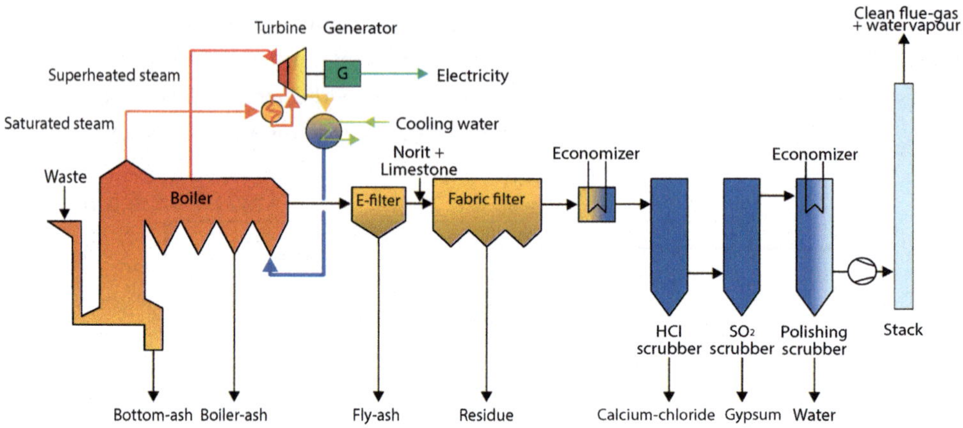

The grate is a water-cooled martin grate W + E licensed. The overall design of the boiler differs from conventional designs for its size. Both empty passes and horizontal passes are substantially bigger. It creates the heating surface necessary for the energy transfer. The flue gas cleaning configuration is a combination of dry and wet cleaning. The first stage is an ESP in combination with a bag filter, the second stage has three scrubbers in a row. Between the bag filter and the quench is a heat recovery step called eco 2 for condensate preheating and in the last scrubber there is an eco 3 also for condensate preheating. The outlet temperature of the flue gas is just under 57°C. The emissions – except for NO_x – are 20% of the emissions allowed by EU IE directive. The NO_x is under 40%. This makes the plant not only the most efficient but also one of the cleanest in the world.

Figure 4 Air emission HRC 2016

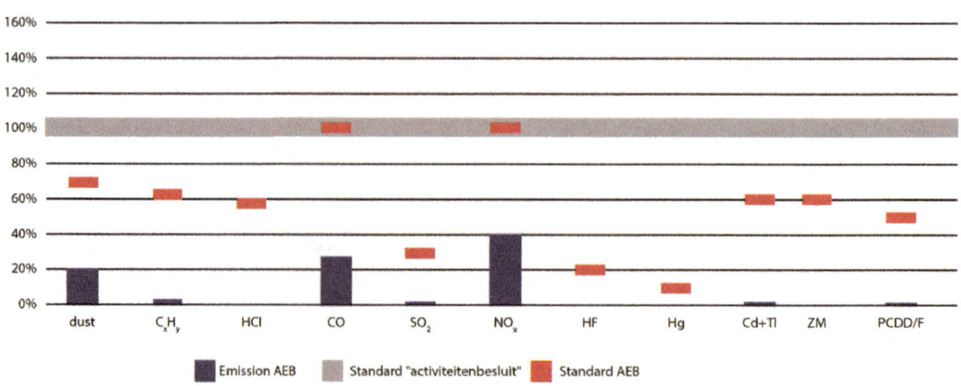

Figure 5 Emissions into air AEC and HRC 2013

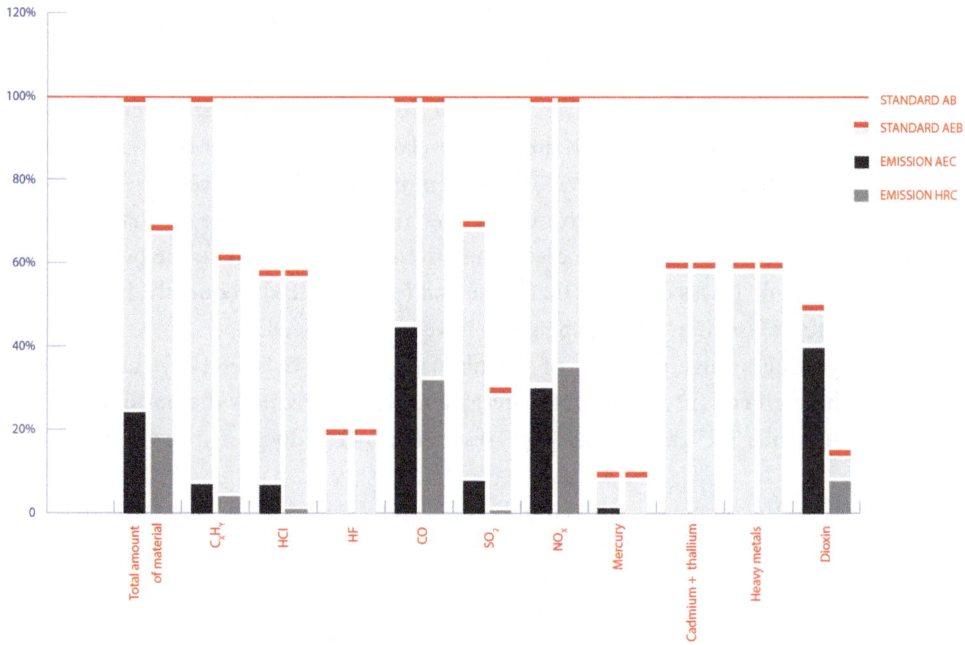

AEC = conventional plant built in 1992 (designed to be clean).
HRC = high-efficiency technology (maximum recovery).

4.7 Viability of High-Efficiency Technology

From a cost perspective, we should always bear in mind that we think in societal costs when it comes to consumer-related waste streams, since specified treatment cost per material is hard to determine in a globalized market. Responsibility programmes for producers are a way to address these costs, although they are differently successful. It means that the costs for treating residual waste streams are generally determined. Other related societal costs are emissions and space occupation. These are all indirect costs, which cannot be related to a traditional business case.

If we look at a traditional business case, we compare capital investments, operational costs and interest. In comparison to a conventional plant, we see an increase in capex of less than 12% to obtain a gain of efficiency from 23 to 30%. This provides an efficiency improvement of 8% on a total investment of €275 (per tonne year). The business case will fly with an electricity price of over €55 per MWh in relation to a gate fee at €65 per tonne MSW. This is at a depreciation time of at least 25 years and at an interest rate of 5%.[12] If governments can put a green fee as an incentive for renewable energy (54% of the MSW

12 Berlo, *et al.*, 2011 (cost model).

has a bio origin), this incentive could lead to lower gate fees in order to find a completive edge with landfill. This is easier to accomplish with high efficiency than with conventional WtE technologies.

The extra 10-12% capex investment for high efficiency creates an increase in electricity income of 25-35%. The gain is even greater if we consider the environmental impact in CO_2 reductions and reward this on the same level as we reward other renewable energy sources. It means that high efficiency >30% will immediately become cost effective compared to conventional WtE plants. From a maintenance perspective we see that the overall costs are equal or less than in the case of a conventional WtE due to the performance of the super-heaters. It was expected that these had to be replaced frequently, but after 10 years of operation less than 10% of the pipes have been replaced.

Landfill produces huge amounts of greenhouse gases. WtE effectively turns a large amount of these greenhouses gases into energy and thereby reduces the burning of "virgin" fossil fuels like oil, gas or coal. The result is a net reduction in greenhouse gas production.

So if we want to compare it with landfill, we need to consider the long-term effect on the environment, especially the greenhouse gas production from landfill.

4.8 Performance and Availability of High-Efficiency Technology during 2010-2016

As reported by Berlo,[13] there were massive problems in the first years of the plant operation during 2007-2009. This was not related to the design but mostly down to bad luck (break of the generator shaft) and poor project management, resulting in poor performances by the contractors. Furthermore, choices in configuration would have been different if budget limitations had not been in force. The operation stabilized in the beginning of 2010 and has remained stable ever since. For this reason, results from the period of 2010-2016 are regarded as representative for the HR technology's performance and potential.

4.8.1 Load Points

Until 2011 and 2012, respectively, the boilers were operating at the so-called 100% – load point, which was within the guarantees of the suppliers. The load was then increased to 110%, by which the viability increased substantially, at the cost of creating a higher risk of corrosion.

13 Berlo, *et al.*, 2011.

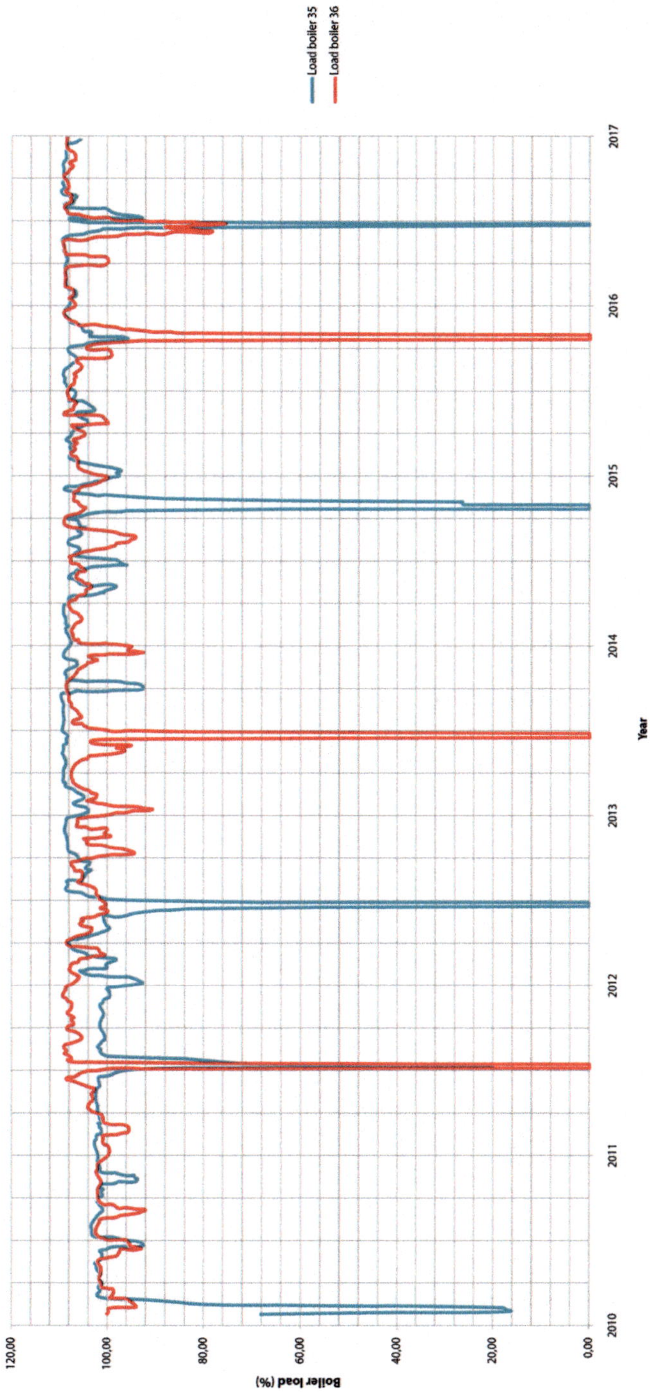
Load boiler 35
Load boiler 36
Boiler load (%)
120,00
100,00
80,00
60,00
40,00
20,00
0,00
2010
2011
2012
2013
2014
2015
2016
2017
Year

The average actual boiler load has been 104.47% during 2010-2016 and 105.84% during 2012-2016 when both boilers were operated at the increased load.

Fluctuation of the boiler load is inherent to the combustion of waste, as the properties of waste as a fuel vary unpredictably within a certain range. Nevertheless, a performance factor of 92% for 2012-2016 is a very good value in WtE. The performance factor is slightly increasing and stable.

4.8.2 *Availability*

This section is focused on the achieved availability of the boilers, in order to identify the down-time caused by mechanisms that are inherent to the HRC-concept. Of course, the overall availability is also strongly influenced by other systems, like the steam turbine. They are not regarded in this report because in terms of technology, the avoidance and control of corrosion at high-efficiency operation levels is the most unique property of the HRC-concept.

4.8.2.1 Boiler Availability

Figure 6 Boiler availability from 2010 to 2016

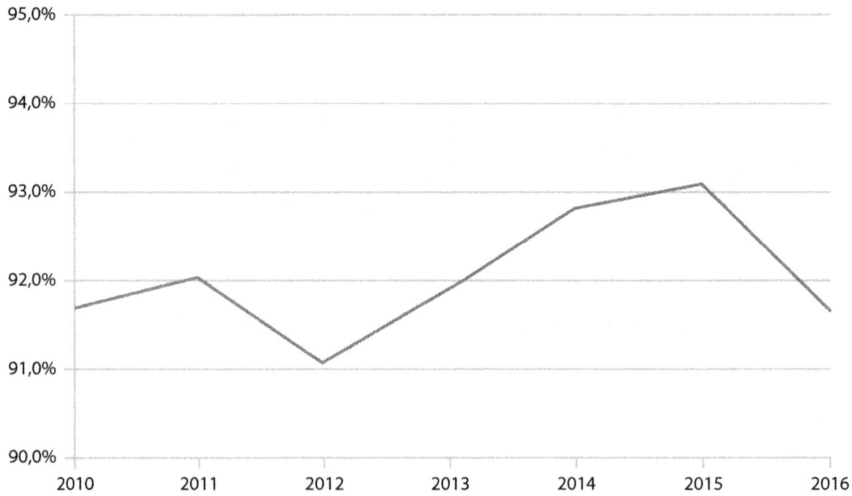

In the whole operational phase from July 2007 to the end of 2016, the total boiler availability of both HRC boilers was on an average 90.9%.

In the period from 2010 until 2016, the boiler availability average was 92.0% with a minimum of 91.1% in 2012 and a maximum of 93.1% in 2015. In 2016, the availability dropped to 91.7%.

It can be stated that from 2010 onwards the availability has been stabilized clearly above 90%.

4.8.2.1.1 Planned versus Unplanned Non-availability

The planned non-availability (green line) is determined by the maintenance schedule. The boilers are shut down once every two years for an overhaul, initially for 21 days. Once every four years the whole plant is shut down for maintenance and inspection of the common parts, which is combined with the two-year planned overhaul of one of the boilers.

For the planned maintenance, delays in the overhauls have been charted and the average down-time was 4.7% for the period of 2010-2016.

The unplanned non-availability (red) decreased rapidly to a 4% level after the first two years of operation. The average unplanned non-availability was 3.3% for the period of 2010-2016 with a minimum of 0.5% in 2015 and a maximum of 4.6% in 2016.

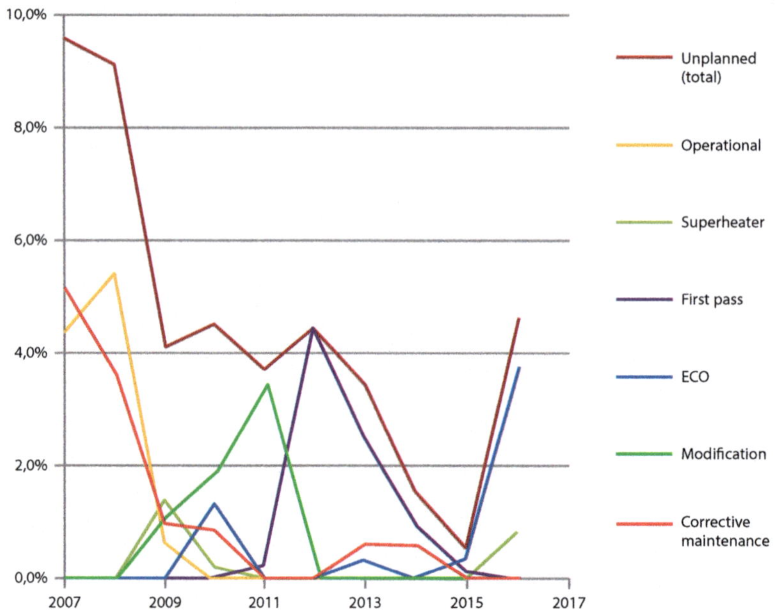

The total unplanned non-availability (dark red) has been split into categories:

- Operational (yellow): Not directly boiler-related causes like de-slagger, residue or fly ash clogging, waste shortage or waste crane disturbance.
- Corrective maintenance (light red): Repair works on miscellaneous equipment, failures like waste cranes, draft fan, flue gas treatment, cooling water system etc.
- Modification (green): Shutdowns needed to modify the process and equipment to avoid the operational and corrective maintenance shutdowns.
- ECO (blue): Leakage of the economizer and its drains.
- Superheater (light green): Leakage of the superheaters and its drains.
- First pass (purple): Leakage of the membrane walls in the first pass due to corrosion.

Following the time line, it is obvious that during 2007-2010 the operation was massively disrupted by operational issues and equipment failures. During 2008-2011, various modifications were implemented to avoid these shutdowns. The effect became visible between 2009 and 2011, when the non-availability from operational issues or corrective maintenance is negligible.

Economizer leakages occurred incidentally in 2010, 2013 and 2015 with a total down-time of 10.49 days. In 2016, damage on the economizers caused 30.2 days of down-time, which is a dramatic increase of these failures. This was caused by a leakage at the centre of the economizer bundle, which took a long time to repair. In the repair, welding works were executed inappropriately, and the leakage recurred twice after several days.

Superheater leakages occurred only incidentally in 2009, 2010 and 2016 with a total down-time of 18 days.

First pass membrane wall leakages occurred from 2011 onwards and are the main cause for unplanned down-time in 2012. The material loss has been noticed in an earlier phase and a massive effort has been made to understand and prevent the failures. The conclusion from the investigation is that this corrosion effect is directly related to the parameters of the HRC-concept design. Equipment design strategies and material quality specifications have been developed to prolong the membrane wall life time. Parts of the membrane walls have been replaced with improved materials and design panels. As a result, no leakages have occurred since 2014.

4.8.3 *Efficiency*

With the shutdown-days excluded, the average net-gross-efficiency is 30.64% in this period.

The efficiency stays clearly above 30% in normal operation. Periods with a lower efficiency result from a reduced performance in the turbine steam reheating cycle due to failures of the condensate pumps in the reheating cycle.

4.8.4 *Energy Recovery*

According to the EU 2008 Waste Framework Directive (WFD), waste incineration is regarded as recovery if the recovery rate according to its annex R1 is higher than 0.6.

The average R1 recovery rate is 0.92 for 2010-2016. In the graph, it is split into electricity (orange) and heat (red); the average contribution of heat to the recovery rate is 3.2% only.

The energy production has been calculated as fictive energy efficiency and normalized to standard waste with a heat content of 10 MJ/kg.

Repeatedly, a monthly average higher than 1,000 kWh/t has been reached.

Figure 7 Net gross efficiency from 2010 to 2016[14]

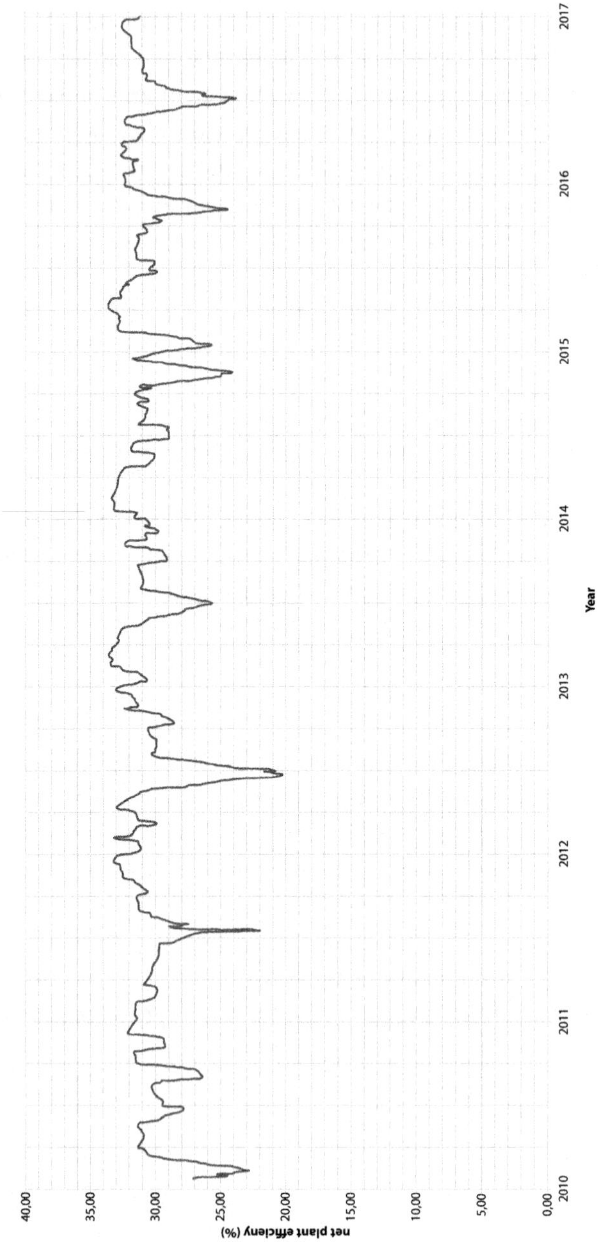

14 30-Day-moving-averages on day-base. Values beneath 10% due to shutdowns have not been taken into account.

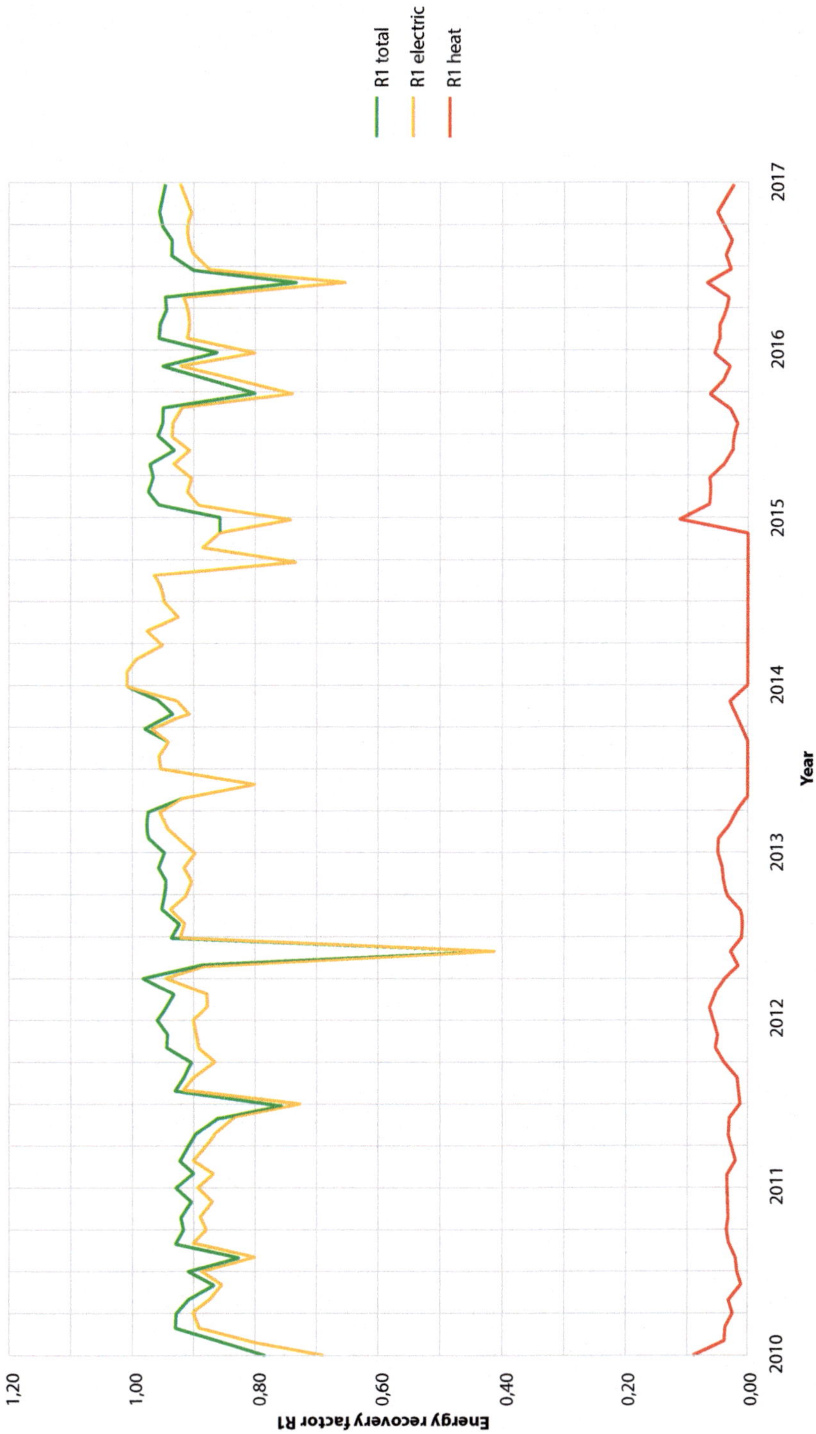
R1 total
R1 electric
R1 heat
Energy recovery factor R1
1,20
1,00
0,80
0,60
0,40
0,20
0,00
Year
2010
2011
2012
2013
2014
2015
2016
2017

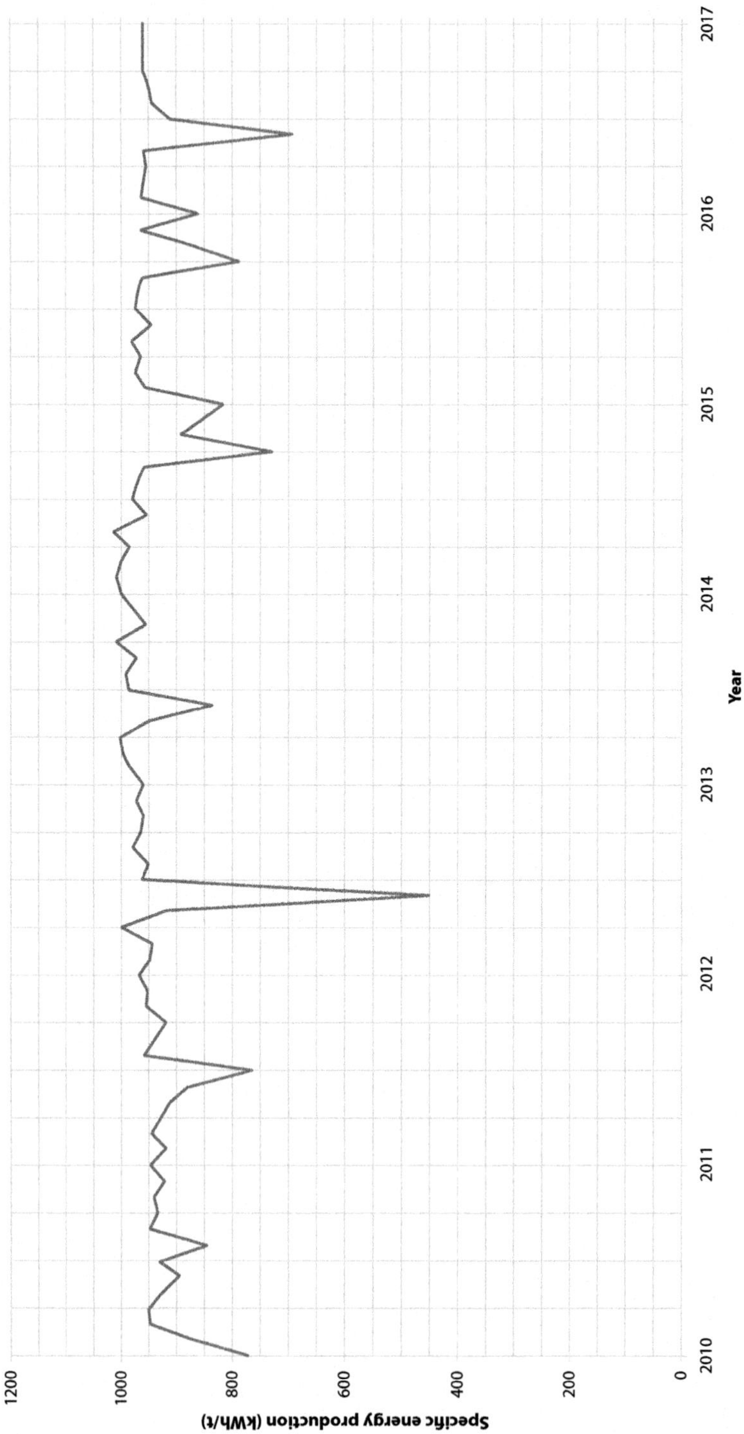
Specific energy production (kWh/t)
1200
1000
800
600
400
200
0
Year
2010
2011
2012
2013
2014
2015
2016
2017

4.9 Future Perspective of the Fifth Generation High-Efficiency WtE Plant

4.9.1 *Future Perspective*

In a recent Shell scenario study,[15] WtE is shown as an important way of generating energy, which can be used for base load production and balancing wind and solar energy production.

If a WtE plant runs in a CHP mode where they can be flexible in the production of heat and electricity, it can be used to balance the grid for unpredictable Wind and solar power.

Also, CCUS (carbon capture utilization and storage) possibilities make WtE a true sink for CO_2 in particular, if we consider that 54% of the input has a short cycle origin in biomass.

At this moment, the CCU (Carbon Capture Utilisation) for greenhouses is under business development and is expected to be in operation around 2021. At that time, together with the port of Amsterdam 50 Kilotonne of CO_2 will be captured and used in greenhouses as a fertilizer.

Another noteworthy development is the emergence of a renewable energy market (local smart-grid) for the port of Amsterdam where AEB will develop a marketplace for producers of renewable energy, wind solar biomass related, and for local end-users of this energy.

The combination of off-peak power transferred into hydrogen and CO_2 capturing makes it possible to create methanol as a basic chemical or renewable fuel for transport.[16] Possibilities for a steam network providing process steam for the industry are already well-known and applied at other WtE facilities in the world and currently under development at AEB. The company runs an extensive innovation programme in order to recover materials of high quality, or efficiently (chemically) transform them into valuable commodities. Furthermore, residual waste streams are converted with the highest efficiency possible into several types of energy, adding up to the overall recycling efficiency.

4.9.2 *Current Industrial Symbioses*

AEB is located next to the city's sewage plant where this plant makes use of the low-exergy residual waste heat in order to maintain their digestion process for cleaning the sewage and producing methane. This "bio-gas" is burned in a gas engine to produce electricity and heat. The sewage sludge is burned in the WtE plant.

15 '8 Colours scenario', Shell 2015.

16 According to BSE engineering, this is already feasible at an electricity price below 65 €/MWh.

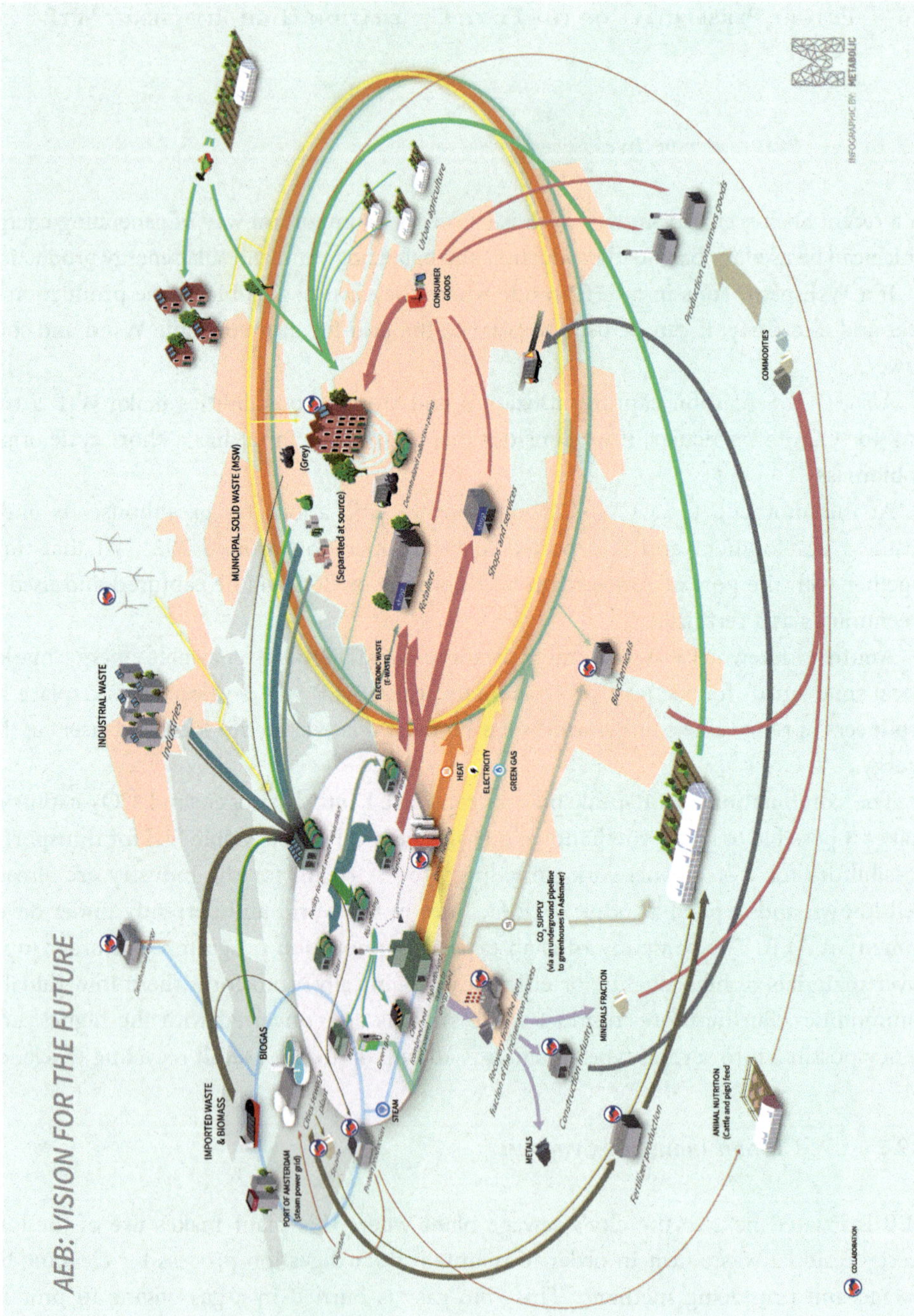

Obviously, if we want to use these options elsewhere, the location of a WtE facility is to be chosen carefully.

4.10 Conclusion

Thermal conversion of residual waste streams using high-efficiency technology will be necessary, also in a more circular economy where recycling is possibly at a maximum. At least through the transition period we are in, which is expected to last a few more decades. High efficiency adds up to recycling providing the best solutions for residual waste streams from the recycling processes.

Technologies that are the most suitable for this conversion should be judged by their environmental impact, efficiency, overall societal costs and sink possibilities. This means that with the right incentives governments can push this technology to the edge.

Often high-efficiency WtE technology will be the starting point for developing the circular economy following the waste hierarchy bottom up, knowing that the sink is in place for residuals from recycling processes and other end-of-life materials.

As a "must run" installation, industrial symbioses can easily be found with complementary industrial processes as a reliable source. While producing CO_2, electricity and heat, the conditions for electrification of the chemical industry are in place. With the development of renewable energy, "must run" can then be used for balancing the grid. The inert fraction produced can serve after a cleaning process for construction purposes as a clean commodity replacing sand, granulates and even cement.

High-efficiency technology utilizes power plant technology to improve the efficiency of the generation of electricity from waste. Dedicated process design, corrosion protection and design methods have been applied to the combustion system and the water–steam cycle to achieve the maximum electrical efficiency while keeping the typical corrosion effects under control.

The boilers are operated at 110% load since 2011 and 2012, respectively, with no significant changes in wear and corrosion.

Both set design goals, a gross efficiency higher than 30% and an availability higher than 90% have been met and exceeded year by year since 2010. Inconel625 cladding has shown sufficient corrosion resistance even in critical areas.

Performance figures show that the AEB High-efficiency-technology is operating at the desired high efficiency in a very reliable manner. The availability figures show that the applied preventive maintenance schedule and unplanned corrective maintenance are in the range of conventional WtE plants.

With its extensive and sophisticated flue gas cleaning path, the emissions of the plant score better or equal than any other WtE facility in the world. Almost at every level, less than 20% of the allowed emissions set by the Industrial Emissions Directive are achieved. The extensive bio-monitoring programme shows that no harm is done to the crops and cow milk produced in the vicinity of the plant.

The High-efficiency-technology maximizes the CO_2 reduction in waste treatment as it achieves very high R1 values from the generation of electricity alone. These high recovery factors can also be achieved if process steam or heat delivery cannot be applied in the surroundings of the plant (like in southern Europe). Meanwhile, locations should be carefully chosen: preferably in the vicinity of other process industries, chemical industries and cities' sewage plants in order to gain maximum industrial symbioses with these industries.

On the basis of the performance, reliability and emission figures achieved over a period of 7 years, the conclusion can be drawn that the AEB High-efficiency-technology is a proven technology, making it a clean and safe sink solution for the circular economy now and in the future. Therefore, this technology should be considered as Best Available Technology (BAT).

References

Chris van Dijk, Wim van Doorn, Bert van Alfen, 'Long term plant biomonitoring in the vicinity of waste incinerators', in The Netherlands, Wageningen, 2015 ('Van Dijk, *et al.*, 2015').

Marcel van Berlo, Harry de Waart, 'Unleashing the power in waste; Comparison of Greenhouse gas and other Performance Indicators for Waste-to-Energy concepts and Landfilling', AEB, Amsterdam, 2008 ('Berlo, *et al.*, 2008').

Martin J. Murer, Hartmut Spliethoff, Chantal M.W. de Waal, Saskia Wilpshaar, Bart Berkhout, Marcel A.J. van Berlo, Oliver Gohlke, Johannes J.E. Martin, 'High Efficient Waste-to-Energy in Amsterdam – Getting ready for the next steps', AEB, Amsterdam, 2011 ('Berlo, *et al.*, 2011').

Shell scenario's 8 colours 2015.

Shell scenario's: chapter; 'a tale of two Dams' 2017, <www.shell.com/energy-and-innovation/the-energy-future/scenarios/new-lenses-on-futurecities/_jcr_content/par/textimage_1687505569.stream/1447854282580/c391a74670d29b3e8f4 f64a70a6d5653fb1f9fbeef0e-de22dd2daccdb5cdab2c/newlensesonfuturecities-june-2014.pdf>.

Thomas Rau and Sabine Oberhuber, Materials Matter, Bertram & De Leeuw, Haarlem, 2016.

Websites

<www.aebAmsterdam.nl>

<www.eia.org>

<https://www.ellenmacarthurfoundation.org>

5 Assessing WtE Technologies by Way of an Analytic Hierarchy Process

Filippo Corsini, Natalia Gusmerotti and Marco Frey

Nowadays, sustainability within the waste management system is a hot topic. The EU Communication (COM (2015) 614 final), "Closing the Loop – An EU Action Plan for the Circular Economy", highlights that when waste cannot be prevented or recycled, recovering its energy content is in most cases preferable to landfilling it, in both environmental and economic terms. For the EU Commission "Waste to Energy" can play a role and create synergies with EU energy and climate policy, but only if guided by the principles of the EU waste hierarchy. Therefore, one of the main challenges is how to optimize its role without compromising on the achievement of higher reuse and recycling rates, and how to best exploit the corresponding energy potential. However, despite the various researches available in the field of waste valorization, a large number of issues remain unaddressed. One of the main issues is the assessment of benefits, economic opportunities, costs and risks of waste to energy processes. A number of waste-to-energy (WtE) technologies are currently proposed and practised. Assessing their potential (in terms of benefits, economic opportunities, costs and risks) is now possible by gathering the existing operating experience and the data available.

The chapter uses in-depth literature review of research undertaken in this field to retrieve currently available data. Then, after defining a conceptual framework to classify WtE technologies, a multi-criteria decision-making methodology has been applied to evaluate the trade-offs between the benefits, economic opportunities, costs and risks of different WtE alternatives.

To develop a credible comparison between alternatives, four academic experts in the field of WtE have been engaged and their opinions gathered, using surveys. Results and conclusions allow us to better understand the contribution WtE processes can make to reach the goals of waste management in a circular economy perspective.

5.1 Introduction

In the last few years, the concept of circular economy (CE) is attracting more and more attention of policy makers, businesses, practitioners and the scientific community. An industrial system is considered circular if it is restorative or regenerative by intention and design.[1] Such a system represents an alternative to the "linear economy", where the industrial process is characterized by flows of raw materials that are transformed into a product and, after the use phase, into a disposable waste.[2] The CE aims to optimize the use of natural resources by increasing the efficiency towards closed cycles of materials and energy and industrial processes that generate less amounts of waste. This requires identifying alternative solutions based on the life cycle of a process and on the interaction between such a process and the environment and the economy in which it is embedded.[3]

Many authors[4] refer here to different schools of thought, such as Pearce and Turner[5] who elaborated the concept of a circular economy system taking inspiration from the previous study conducted by Boulding,[6] the General Systems Theory of Von Bertanlaffy[7] and Industrial Ecology.[8] Furthermore, the Ellen Macarthur Foundation[9] links the CE concept to more recent theories such as regenerative design, performance economy, cradle to cradle, bio-mimicry and blue economy.

Reduction, Reuse and Recycle principles, which are connected to CE in several studies on this topic,[10] are also included in the waste hierarchy of the EU Waste ("Framework") Directive 2008/98/EC (EC, 2008).[11] Aiming at promoting the further development of the CE approach, the Ellen Macarthur Foundation[12] mixed these principles with other ones: (i) appropriate design, focusing on the design phase of products in order to avoid land-

---

1 Ellen MacArthur Foundation, *'Towards the circular economy: an economic and business rationale for an accelerated transition'*, 2012.

2 V. Elia, M. G. Gnoni, F. Tornese, *'Measuring circular economy strategies through index methods: A critical analysis'*, Journal of Cleaner Production, Vol. 142, 2017, pp. 2741-2751.

3 P. Ghisellini, C. Cialani, S. Ulgiati, *'A review on circular economy: The expected transition to a balanced interplay of environmental and economic systems'*, Journal of Cleaner Production, Vol. 114, 2016, pp. 11-32.

4 *See*, for instance, CIRAIG, *'Circular economy: a critical literature review of concepts'* Montréal, Québec, Canada. Available from <www.ciraig.org/pdf/CIRAIG_Circular_Economy_Literature_Review_Oct2015.pdf>.
See also Ghisellini et al., *supra* note 3.
See also M. Lieder, A. Rashid *'Towards circular economy implementation: A comprehensive review in context of manufacturing industry'*, Journal of Cleaner production, Vol. 115, 2016, pp. 36-51.

5 D.W. Pearce, R.K. Turner, *Economics of Natural Resources and the Environment*, Hemel Hempstead, Harvester Wheatsheaf, London, 1989.

6 K. Boulding, *'The economy of the coming spaceship earth'*, In: Daly, H., *Freeman, Economics, Ecology, Ethics: Essay towards a Steady State Economy*, San Francisco, 1966.

7 L. Von Bertalanffy, *General System Theory*, New York, George Braziller, 1968.

8 F. Preston, *'A Global Redesign? Shaping the Circular Economy'*, Briefing Paper. Available: <https://www.chathamhouse.org/publications/papers/view/182376> (accessed 10.10.16).

9 Ellen MacArthur Foundation, *supra* note 1.

10 *See*, for example, L. Reh, *'Process engineering in circular economy'*, Particuology Vol. 11, 2013, pp. 119-133.
See also B. Su, A. Heshmati, Y. Geng, X. Yu, *'A review of the circular economy in China: moving from rhetoric to implementation'*, Journal of Cleaner Production, Vol. 42, 2013, pp. 215-277.
See also L.A. Lett, *'Las amenazas globales, el reciclaje de residuos y el concepto de economia circular'*, Rivista Argentina Microbiologia, Vol. 46, No. 1, 2014, pp. 1-2.

11 Directive 2008/98/EC OF THE EUROPEAN PARLIAMENT AND OF THE COUNCIL of 19 November 2008 on waste and repealing certain Directives, in OJ, L 312/3 of 22 Nov. 2008.

12 Ellen MacArthur Foundation, *supra* note 1.

filling when these products become waste; (ii) reclassification of the materials into "technical" (as metals and plastics) and "nutrients" in order to guarantee – respectively – their reuse and their return to the biosphere; (iii) renewability, stressing the role of renewable energy as the main energy source for circular economy. Therefore, in the CE perspective it is possible to close loops both recovering materials and energy from waste, maximizing waste use, minimizing the use of raw materials and the environmental impact of processes.[13] In a circular economy system, WtE technologies can address problems related to energy demand, waste management and greenhouse gas (GHG) emissions.[14]

The importance of a CE is gradually increasing in European policy, as demonstrated, for instance, by the Manifesto of resource efficiency,[15] and more recently by two Communications of European Commission. The first one concerns a zero waste programme for Europe[16] and the second one proposes an European Action Plan for the Circular Economy.[17]

As highlighted by the European Action Plan for the Circular Economy,[18] if waste cannot be prevented or recycled, recovering energy form waste is much better than sending it to a landfill, both from an environmental and from an economic perspective. For the European Commission, WtE, if managed in the light of the EU waste hierarchy, can play a fundamental role since it links energy policy and climate policy. The challenge is how the role of WTE can be optimized without compromising European objectives of a high rate of reuse and recycling of products and waste.

On the basis of the waste hierarchy[19] prevention is the most desirable option as indicated in the "mandatory" five-level hierarchy, followed by preparing for reuse, recycling, other recovery (e.g. energy recovery) and disposal as the last option.[20] Although waste prevention has become a priority worldwide,[21] statistics reveals that only the typi-

13 Ghisellini et al., *supra* note 3.

14 *See*, for instance, S.Y. Pan, M.A. Du, I.T. Huang, I.H. Liu, E.E. Chang, P.C. Chiang, *'Strategies on implementation of waste-to-energy (WTE) supply chain for circular economy system: a review'*, Journal of Cleaner Production, Vol. 108, 2015, pp. 409-421.

15 European Commission, 17/12/2012. Manifesto for a Resource-efficient Europe. MEMO 12/989/, available: <http://europa.eu/rapid/press-release_MEMO-12-989_en.htm> (accessed 25.10.16).

16 European Commission, Communication from the Commission to the European Parliament, the Council, the European Economic and Social Committee and the Committee of the Regions, Towards a circular economy: A zero waste programme for Europe, COM(2014) 398 final, Brussels, 2014.

17 European Commission, Communication from the Commission to the European Parliament, the Council, the European Economic and Social Committee and the Committee of the Regions, Closing the loop – An EU action plan for the Circular Economy, COM(2015) 614 final, Brussels, 2015.

18 *Id.*

19 Directive 2008/98/EC *supra* note 11. Article 4.

20 *See* A.A. Zorpas, K. Lasaridi, *'Measuring waste prevention'*, Waste management, Vol. 33, No. 5, 2013, pp. 1047-1056.
See also U. Gelbmann, B. Hammerl, *'Integrative re-use systems as innovative business models for devising sustainable product–service-systems'*, Journal of Cleaner Production, Vol. 97, 2015, pp. 50-60.

21 *See* J. Cox, S. Giorgi, V. Sharp, *'Household waste prevention – A review of evidence'*, Waste Management and Research, Vol. 28, 2010, pp. 193-219.
See also V. Sharp, S. Giorgi, D. Wilson, *'Methods to monitor and evaluate household waste prevention'*, Waste Management & Research, Vol. 28, No. 3, 2010, pp. 269-280.

cal treatment option (recycling, incineration and landfill) are taken into account by technical reports.[22]

Other forms of recovery, that are not mentioned in the definition section of the EC directive, are all operations meeting the definition for "recovery" but failing to comply with the specific requirements of preparation for reuse or recycling.[23] Recovery is defined by the EC directive as any operation the principal result of which is that waste serves a useful purpose by replacing other materials that would otherwise be used to fulfil a particular function.

As the objective is to exploit the potential of "waste to energy", the EC will launch a specific initiative that will be part of the Energy Union Framework Strategy[24] and of the EU action plan for the circular economy.[25] This initiative is supported by the 7th Environment Action Programme, which requires that measures be taken to ensure that energy recovery is limited to non-recyclable waste.[26]

The forthcoming initiative[27] is expected to address the following problems: (i) assessing the role of WtE under the waste hierarchy as the best option for the management of combustible non-recyclable waste, highlighting most efficient processes for each main waste stream; (ii) indicating under what circumstances it is justified to depart from the waste hierarchy for specific waste streams; (iii) analysing the contribution of WtE as a safe option to eliminate waste containing hazardous substances that should be safely managed so as not to pose any significant risks to human health and the environment.

The scope of our study is to contribute to the ongoing discussion on the feasibility of different technologies for energy generation from municipal solid waste (MSW) in European countries in the light of circular economy objectives using an analytic hierarchy process (AHP). More in detail, this research contributes to a better assessment of the following technologies:

- co-combustion in coal-fired power plant,
- waste incineration with energy recovery,

22 *See* A. Bartl, *'Moving from recycling to waste prevention: a review of barriers and enables'*, Waste Management & Research, Vol. 32, No. 9, 2014, pp. 3-18.
See also EUROSTAT, *'Waste statistics'*, available from <http://ec.europa.eu/eurostat/statistics-explained/index.php/Waste_statistics>, 2015.

23 European Commission, Guidance on the interpretation of key provisions of Directive 2008/98/EC on waste, Brussels, 2012.

24 European Commission, Communication from the Commission to the European Parliament, the Council, the European Economic and Social Committee and the Committee of the Regions and the European Investment Bank, A Framework Strategy for a Resilient Energy Union with a Forward-Looking Climate Change Policy, COM(2015) 80 final, Brussels, 2015.

25 European Commission, *supra* note 17.

26 European Union, Decision n. 1386/2013/EU of the European Parliament and of the Council of 20 November 2013 on a General Union Environment Action Programme to 2020 'Living well, within the limits of our planet', Brussels, 2013.

27 European Commission, Exploiting the Potential of Waste to Energy under the Energy Union Framework Strategy and the Circular Economy, Brussels, 2016.

- gasification,
- biogas production from anaerobic digestion.

We will examine the feasibility of these technologies in three European countries, namely Denmark, the United Kingdom and Italy.

The rest of the study is structured as follows. Section 5.2 presents the state of the art in the use of Multi-Criteria Analysis for supporting decision-making regarding waste management. In Section 5.3, applied methodologies and data collection processes are described. In Section 5.4, we present the results obtained. The last section provides the main conclusions, as well as a discussion of their implications, limitations and ways forward in further studies.

5.2 MULTI-CRITERIA ANALYSIS AND WASTE MANAGEMENT

The MSW management decision-making process experienced substantial changes in the last 30 years.[28] As the concept of sustainability became central to waste management, the use of tools able to consider different criteria (i.e. economic, environmental, technological, social, etc.) quickly took root. The most commonly used models are Cost Benefit Analysis, Life Cycle Analysis and Multi-criteria Analysis (MCA). Particularly MCA optimizes the decision-making process as it provides a systematic methodology to rank many concrete alternatives from the best to the worst ones based on multiple criteria.[29] In the waste management field, MCA is frequently used to tackle numerous issues (selection of the sites of facilities, selection of waste management options, enhancing public participation, etc.). In fact, although a large number of criteria can complicate decision-making, the complexity of waste management systems justifies this approach.[30]

Due to the possibility of leading to more open and accountable decision-making processes, the potential of MCA as a public participation tool is amply recognized.[31] MCA, in

---

28 J.P. Su, P. Chiueh, M. Hung, H. Ma, '*Analysing policy impact potential for municipal solid waste management decision-making: A case study of Taiwan*', Resources, Conservation and Recycling, Vol. 51, 2007, pp. 418-434. *See also* Chapter 3 by Helsen *et al.*, in this volume.

29 M. Behzadian, R.B. Kazemzadeh, A. Albadvi, M. Aghdasi, '*PROMETHEE: A comprehensive literature review on methodologies and applications*', European Journal of Operational Research, Vol. 200, 2010, pp. 198-215.

30 R. Oyoo, R. Leemans, A. Mol, '*The determination of an optimal waste management scenario for Kampala, Uganda*', Waste Management & Research, Vol. 31, No. 12, 2014, pp. 1203-1216.

31 *See* H. Hu, L. Xiang, A.D. Nguyen, P. Kavan, '*A critical evaluation of waste incineration plants in Wuhan (China) based on site selection, environmental influence public health and public participation*', International Journal of Environmental Research and Public Hearth, Vol. 12, 2015, pp. 7593-7614.
See also M.L. Hung, H. Ma, W.F. Yang, '*A novel sustainable decision making model for municipal solid waste management*', Waste Management, Vol. 27, 2007, pp. 209-219.
See also M.F. Norese, '*ELECTRE III as a support for participatory decision-making on the localisation of waste-treatment plants*', Land Use Policy, Vol. 23, 2006, pp. 76-85.

fact, is a useful technique for engaging local stakeholders in waste management planning.[32]

MCA has been used in evaluating specific issues connected with waste management. Important examples are: the different options for energy recovery from municipal solid waste in both developed and developing countries,[33] the management of specific typologies of waste[34] and construction and demolition waste.[35]

5.3 Materials and Methods

5.3.1 *Data Collection*

The model for the assessment of the feasibility of four technologies for energy generation from MSW is implemented using data collected from the scientific literature. The literature was explored using highly specific search terms. All selected documents were read independently and iteratively discussed by the three researchers involved in the project, thereby ensuring reliability through consensus and minimization of bias. A worksheet was designed in order to collect the specific information from each source. Finally, summary tables of the obtained data were constructed.

5.3.2 *AHP Methodology*

AHP, developed by Saaty,[36] allows dividing the decision problems into understandable parts; each of these parts is analysed separately and integrated in a logical manner. It is a comprehensive methodology designed to facilitate the decision-making process by using

32 *See* D. Hanan, S. Burnley, D. Cook, *'A multi-criteria decision analysis assessment of waste paper management options'*, Waste Management, Vol. 33, 2013, pp. 566-573.
See also F. Contreras, K. Hanaki, T. Aramaki, S. Connors, *'Application of analytical hierarchy process to analyze stakeholders preferences for municipal solid waste management plans, Boston, USA'*, Resources, Conservation and Recycling, Vol. 52, 2008, pp. 979-991.

33 D.J. Nixon, P.K. Dey, S.K. Ghosh, P.A. Davies, *'Evaluation of options for energy recovery from municipal solid waste in India using the hierarchical analytical network process'*, Energy, Vol. 59, 2013, pp. 215-223.
See also L. Kondili, I. Bartzis, K. Kapsali, J.K. Kaldellis, *'Multicriteria analysis of municipal solid waste energy recovery technologies in Greece'*, Global NEST Journal, Vol. 14, No. 2, 2012, pp. 225-234.

34 *See* J. Pabule, A. Blumberga, F. Romagnoli, D. Blumberga, *'Finding an optimal solution for biowaste management in the Baltic States'*, Journal of Cleaner Production, Vol. 88, 2015, pp. 214-223.
See also A. El Hanandeh, A. El-Zein, *'The development and application of multi-criteria decision-making tool with consideration of uncertainty: The selection of a management strategy for the bio-degradable fraction in the municipal solid waste'*, Bioresource Technology, Vol. 101, 2010, pp. 555-561.

35 *See* N. Roussa, C. Dujet, J. Mehu, *'Choosing a sustainable demolition waste management strategy using multicriteria decision analysis'*, Waste Management, Vol. 29, 2009, pp. 12-20.

36 T.L. Saaty, *'An exposition of the AHP in reply to the paper: Remarks on the analytic hierarchy process'*, Management science, Vol. 36, No. 3, 1980, pp. 259-268.

both empirical data and subjective judgments and it allows the deduction of a scale of priorities that is valid to make complex decisions. Such a method has been used for solving complex decision-making problems in various disciplines and currently is widely used in the field of waste management.[37]

According to the axioms on which the AHP methodology is based,[38] we structured the goal of the identification of the feasibility of four technologies for energy generation from MSW. In particular, we structured the decision hierarchy according to the goal to be attained in the following order: the objectives from a broad perspective, through the intermediate levels (criteria on which subsequent elements depend) up to the lowest level (a set of the alternatives). According to Figure 1, the hierarchical structure of the decision-making consists of the following levels: the first level represents the ultimate goal of the decision hierarchy, the second level introduces the criteria, the third level shows the sub-criteria that are utilized in this work and, finally, the fourth level shows the alternatives.

Figure 1 Hierarchical structure

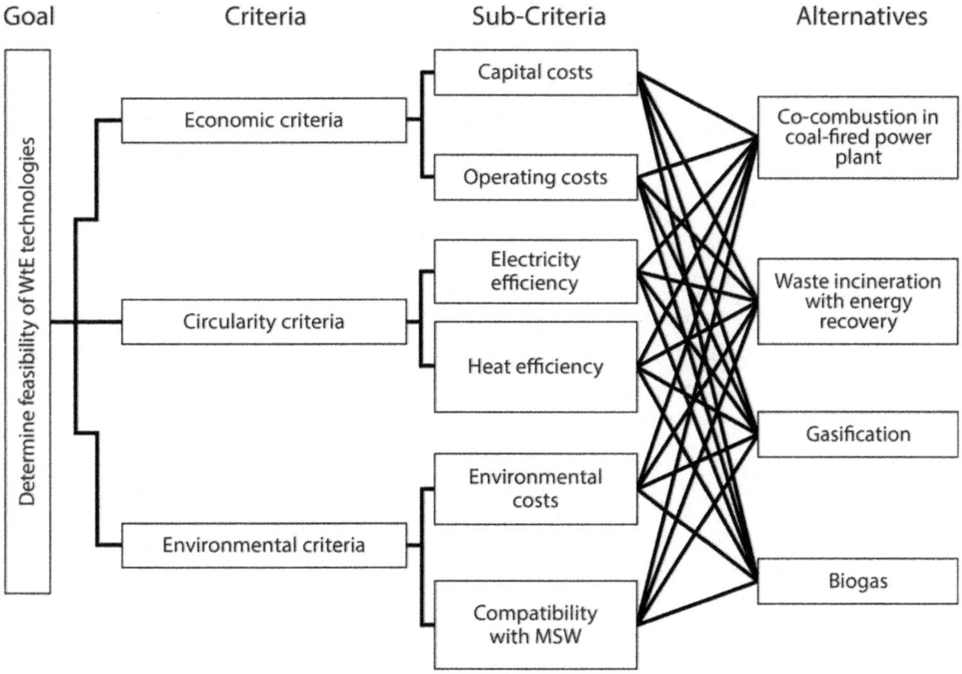

37 *See* A. Soltani, K. Hewage, B. Reza, R. Sadiq, *'Multiple stakeholders in multi-criteria decision-making in the context of Municipal Solid Waste Management: A review'*, Waste Management, Vol. 35, 2015, pp. 318-328.

38 *See* L.G. Vargas, *'An overview of the analytic hierarchy process and its applications'*, European Journal of Operational Research, Vol. 48, No. 1, 1990, pp. 2-8.

5.3.3 *Technologies Assessed*

The WtE technologies compared in this chapter are presented in the following paragraphs. The technologies[39] are selected with the aim of representing a wide spectrum of possible alternatives. They also have different outputs and characteristics in terms of flexibility, which makes it possible to analyse the importance of these technologies for the energy system.

5.3.3.1 Co-combustion in a Coal-fired Power Plant

Co-combustion in a coal-fired power plant represents a process where mixed municipal waste and production-specific commercial waste are co-combusted in coal-fired plants. Increasing power supply from renewable sources of energy in part influences the co-combustion activities. Nowadays, also the trends of CO_2 allowances and the prices of solid recovered fuels as well as imported coal have a significant influence.[40]

5.3.3.2 Waste Incineration with Energy Recovery

Waste incineration with energy recovery is the most common WtE implementation. Modern plants in OECD countries incinerating waste must meet strict emission standards and are vastly different from old types, some of which recovered neither energy nor materials.[41]

5.3.3.3 Gasification

Gasification is a thermochemical treatment process that involves the partial oxidation of a substance. Nowadays, gasification is a well-established technology; however, there are a limited number of commercial plants processing MSW.[42] The main product from gasification is a syngas that can be combusted to produce energy and heat.[43]

---

39 For a detailed explanation of WtE technologies, *see* Chapter 3 by Helsen *et al.*, in this volume.

40 S. Thiel, K.J. Thomé-Kozmiensky, *'Co-combustion of solid recovered fuels in coal-fired power plants'*, Waste Management & Research, Vol. 30, No. 4, 2012, pp. 392-403.

41 H. Rylander, J. Haukohl, *'Status of WTE in Europe'*, Waste Management World, International Solid Wastes Association, 2002.

42 N. Tanigaki, Y. Ishida, M. Osada, *'A case-study of landfill minimization and material recovery via waste co-gasification in a new waste management scheme'*, Waste Management, Vol. 37, 2015, pp. 137-146.

43 H.Y. Yap, J.D. Nixon, *'A multi-criteria analysis of options for energy recovery from municipal solid waste in India and the UK'*, Waste Management, Vol. 46, 2015, pp. 265-277. See also Chapter 7 in this book.

5.3.3.4 Biogas Production from Anaerobic Digestion

Anaerobic digestion is a biological conversion process, which is carried out in the absence of an electron acceptor such as oxygen.[44] The main products of this process are a digest residue and an energy-rich biogas.[45]

5.3.4 *Criteria Selection and Description*

Most of the sub-criteria for assessing the scenarios were selected after an in-depth analysis of the literature regarding the criteria used by other authors in similar studies dealing with the evaluation of waste management scenarios. The six sub-criteria selected are: economic criteria (i.e. capital costs and operational costs), circularity criteria (i.e. electricity efficiency and heat efficiency) and environmental criteria (environmental costs and compatibility with MSW). The following sections give a short presentation of the chosen indicators.

5.3.4.1 Economic Criteria

The economic aspects are addressed through two basic criteria: capital and operational costs. Capital costs represent the costs for the construction of the waste management plant; such costs include: land costs, the costs of design and construction of the plant, and the costs of facilities needed for the operation of the plant. In our research, capital costs differ among technologies in each state and are presented in Table 1.

Operational costs are represented by the expenditures needed during the operation and maintenance of the considered alternatives.[46] Also, in this case, costs differ per technology in each state and are presented in Table 1. Capital costs and operational costs were mainly derived from previous researches.[47]

---

44 I. Angelidaki, D.J. Batstone, *'Anaerobic digestion: process'*, Solid Waste Technology & Management, Vol. 1, 2010, pp. 583-600.

45 H.K. Ahn, M.C. Smith, S.L. Kondrad, J.W. White, *'Evaluation of biogas production potential by dry anaerobic digestion of switchgrass–animal manure mixtures'*, Applied Biochemistry and Biotechnology, Vol. 160, No. 4, 2010, pp. 965-975.

46 Pabule et al., *supra* note 34.

47 *See* M. Münster, H. Lund, *'Comparing waste-to-energy technologies by applying energy system analysis'*, Waste Management, Vol. 30, No. 7, 2010, pp. 1251-1263.
See also Yap & Nixon, *supra* note 43.
See also V. Iaboni, P. Stefanis, *'Aspetti economici del recupero energetico da rifiuti urbani'*, ENEA, 2007 (<www. enea. it/com/web/pubblicazioni/volumi.html>).
See also R. Marchesi, P. Bombarda, A. Casalegno, L. Colombo, A. Rota, *'Costi di produzione di energia elettrica da fonti rinnovabili'*. Report ordered by aeeg to politecnico di milano-dipartimento di energia, Politecnico di Milano-Dipartimento di Energia, 2010.

Table 1 Capital costs and operational costs

	Denmark		United Kingdom		Italy	
	Capital costs (M€/ MW)	Operational costs (% on capital cost)	Capital costs (M€/ MW)	Operational costs (% on capital cost)	Capital costs (M€/ MW)	Operational costs (% on capital cost)
Co-combustion in coal-fired power plant	1.2	10%	1.1	10%	0.95	10%
Waste incineration with energy recovery	5.5	7%	4.2	10%	4.1	9%
Gasification	5	4%	5.7	11%	5	10%
Biogas production from anaerobic digestion	2.7	7%	2.5	4.5%	1.8	3%

5.3.4.2 Circularity Criteria

The circularity aspects are addressed through two basic criteria: electricity efficiency and heat efficiency. Indeed, we assumed that all the technologies assessed, in order to avoid efficiency loss, are able to also to produce heat power. We estimated the efficiency of each technology based on the previous work conducted by Münster and Lund,[48] Yap and Nixon[49] and Brunner and Rechberger.[50] For these indicators, we assumed the same efficiencies in each of the three countries assessed. Table 2 presents the electricity efficiency and heat efficiency for the four technologies assessed.

Table 2 Electricity efficiency and heat efficiency

	Electricity efficiency (%)	Heat efficiency (%)
Co-combustion in coal-fired power plant	27	26
Waste incineration with energy recovery	25	65
Gasification	42	45
Biogas production from anaerobic digestion	41	50

48 Münster & Lund, *supra* note 47.
49 *See also* Yap & Nixon, *supra* note 43.
50 P.H. Brunner, H. Rechberger, *'Waste to energy – key element for sustainable waste management'*, Waste Management, Vol. 37, 2015, pp. 3-12.

5.3.4.3 Environmental Criteria

The environmental aspects are addressed through two basic criteria: environmental costs and the compatibility with the MSW.

WtE plants emit pollutants such as sulphur dioxide, lead and dioxins that can be linked to environmental problems and health damages if they occur in high concentrations. We considered the damage from emissions in the air (mainly CO_2, NO_x and SO_2) as environmental costs calculated based on the results obtained by Jamasb and Nepal.[51]

Finally, we considered the compatibility with the MSW. This indicator takes into account mainly the composition of MSW in the three countries and has been assessed qualitatively; therefore, the 9-level scale established in the AHP method[52] was used for the assessment of this criterion. We used the following options to evaluate the compatibility with MSW "high compatibility", "medium compatibility" and "low compatibility".

5.3.5 *Pair-wise Comparison*

AHP methodology has been used to weigh criteria, to analyse the collected data and to expedite the decision-making process. In general, AHP is used to establish priorities among the elements within each stratum of the hierarchy. This is done by asking a board of decision-makers to evaluate a set of elements in a pairwise comparison. In our case, four experts working in the Italian waste management sector were consulted. They included scientists working at a university in environmental and economic science as well as experts dealing with the problem of waste management.[53]

In the AHP methodology, the hierarchy is broken down into a series of pair comparison matrices, and the participants are asked to evaluate the relationship with respect to the goal. A 9-point scale, ranging from 1 (indifference or equal importance) to 9 (extreme preference or absolute importance) is generally used. This pairwise comparison enables the decision-maker to evaluate the contribution of each factor to the objective independently, simplifying the decision-making process.[54] In our case, a judgmental matrix, denoted by **A**, has been formed using the comparisons. Each entry a_{ij} of the judgmental matrix is formed comparing the row element $\mathbf{A}_i$ with the column element $\mathbf{A}_j$:

$$\mathbf{A} = (a_{ij})\ (i, j = 1, 2, \ldots)$$

---

51 T. Jamasb, R. Nepal, *'Issues and options in waste management: A social cost–benefit analysis of waste-to-energy in the UK'*, Resources, Conservation and Recycling, Vol. 54, No. 12, 2010, pp. 1341-1352.

52 Saaty, *supra* note 36.

53 More in detail, we interviewed one professor of economics and business management, one researcher in environmental economics and two waste management industry insiders.

54 *See* T. L. Saaty, *'Fundamentals of decision making and priority theory with the analytic hierarchy process'*, Vol. 6, Rws Publications, Pittsburgh, 2000.

The 9-point scale has been used to transform the judgements into numerical quantities representing the values of a_{ij}. All the entries a_{ij} are governed by the following rules:

$$a_{ij} > 0;\ a_{ij} = 1/a_{ji};\ a_{ii} = 1 \text{ for all } i$$

Because of the above rules, the judgmental matrix A resulted in a positive reciprocal pairwise comparison matrix. The weight coefficients of the ranking criteria and the decision sub-criteria have been calculated by finding the principal eigenvector w of the matrix **A**:

$$\mathbf{A}w = \lambda_{\max} w$$

Finally, the grading values of all the criteria were normalized to 1.

In order to test the consistency of the judgment matrix, the consistency ratio coefficient (CR) has been calculated using the methodology of Saaty.[55] In our case, the CR resulted was 0.02. In general, a consistency ratio of 0.1 or less is considered acceptable indicating the overall consistency of the pairwise comparison matrix.[56]

5.4 RESULTS

5.4.1 *Scenario Ranking*

Following the pair-wise comparison, the criteria weight with respect to the goal is shown in Figure 2.

Figure 2 Criteria weight with respect to the goal

As presented in the figure, the environmental criteria were the most important ones followed by electricity efficiency and capital costs. Heat efficiency and operational costs

55 Saaty, *supra* note 36.

56 R. Ramanathan, '*A note on the use of the analytic hierarchy process for environmental impact assessment*', Journal of Environmental Management, Vol. 63, No. 1, 2001, pp. 27-35.

were the ones evaluated as less important in the identification of the feasibility of technologies for energy generation from MSW.

We used the criteria weight with respect to the goal to evaluate the feasibility of technologies for energy generation from MSW in Denmark, the United Kingdom and Italy. Results for Denmark are presented in Figure 3.

As presented in the figure, gasification is the most feasible technology for energy generation from MSW in Denmark. This result is in line with findings presented by Münster and Lund[57] showing that gasification, even if there are a limited number of working plants, is an interesting solution for countries like Denmark.

Due to the low organic content of Danish MSW, biogas production from anaerobic digestion scored third out of the four technologies assessed. Finally, co-combustion in coal-fired power plants is the less feasible solution mainly due to the high environmental costs associated with this technology. Results for the United Kingdom are presented in Figure 4.

For the United Kingdom, a different situation appears. Indeed, the four technologies for energy generation from MSW scored almost on the same level. Waste incineration with energy recovery is the most feasible solution followed by gasification and biogas production from anaerobic digestion. Also, in this case, co-combustion in coal-fired power plants turned out to be the less feasible solution mainly due to the high environmental costs associated with this technology. Results for Italy are presented in Figure 5.

In Italy, biogas production from anaerobic digestion is the most feasible solution. This result is mainly explained by the high compatibility of the technology with the high organic content of MSW. Waste incineration with energy recovery is the second most feasible solution followed by gasification. Finally, also in this case, co-combustion in coal-fired power plant turned out to be the less feasible solution.

57 Münster & Lund, *supra* note 47.

Figure 3 Technology comparison in Denmark

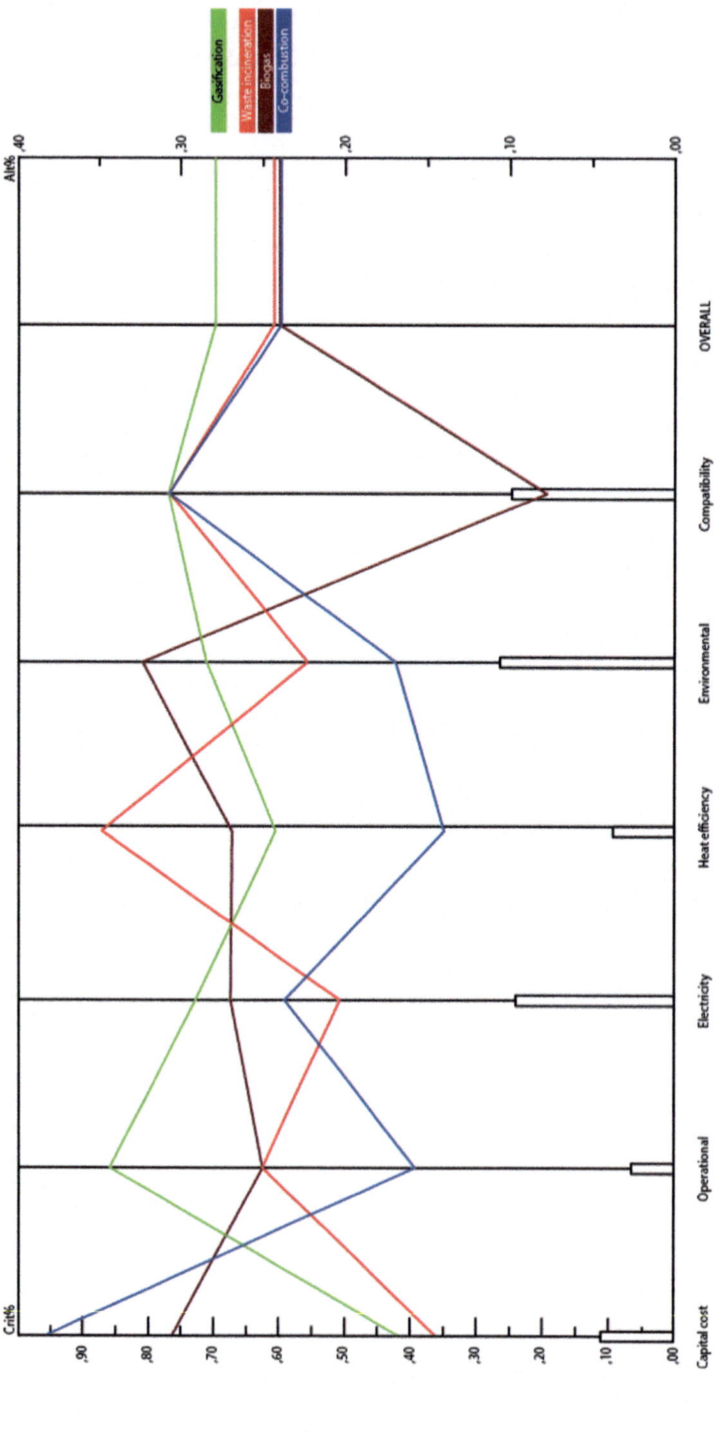

Figure 4 Technology comparison in United Kingdom

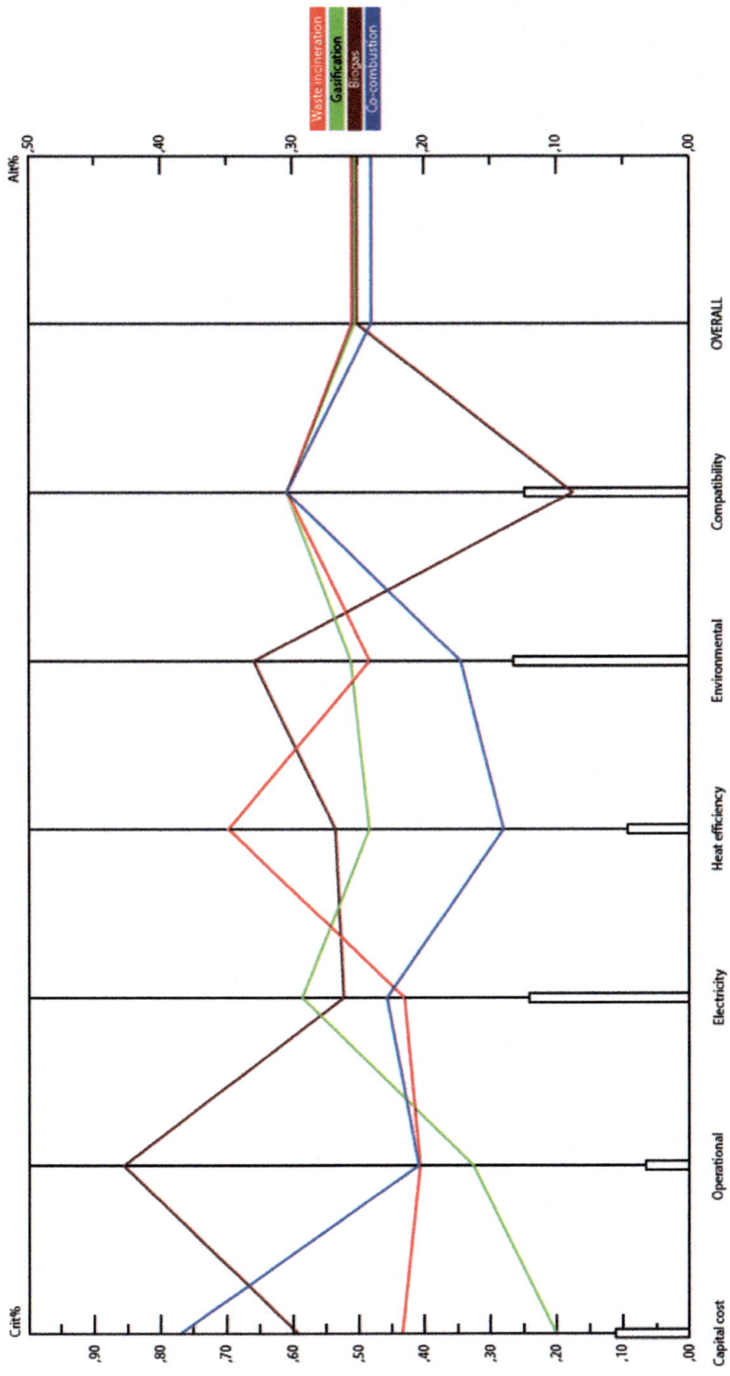

Figure 5 Technology comparison in Italy

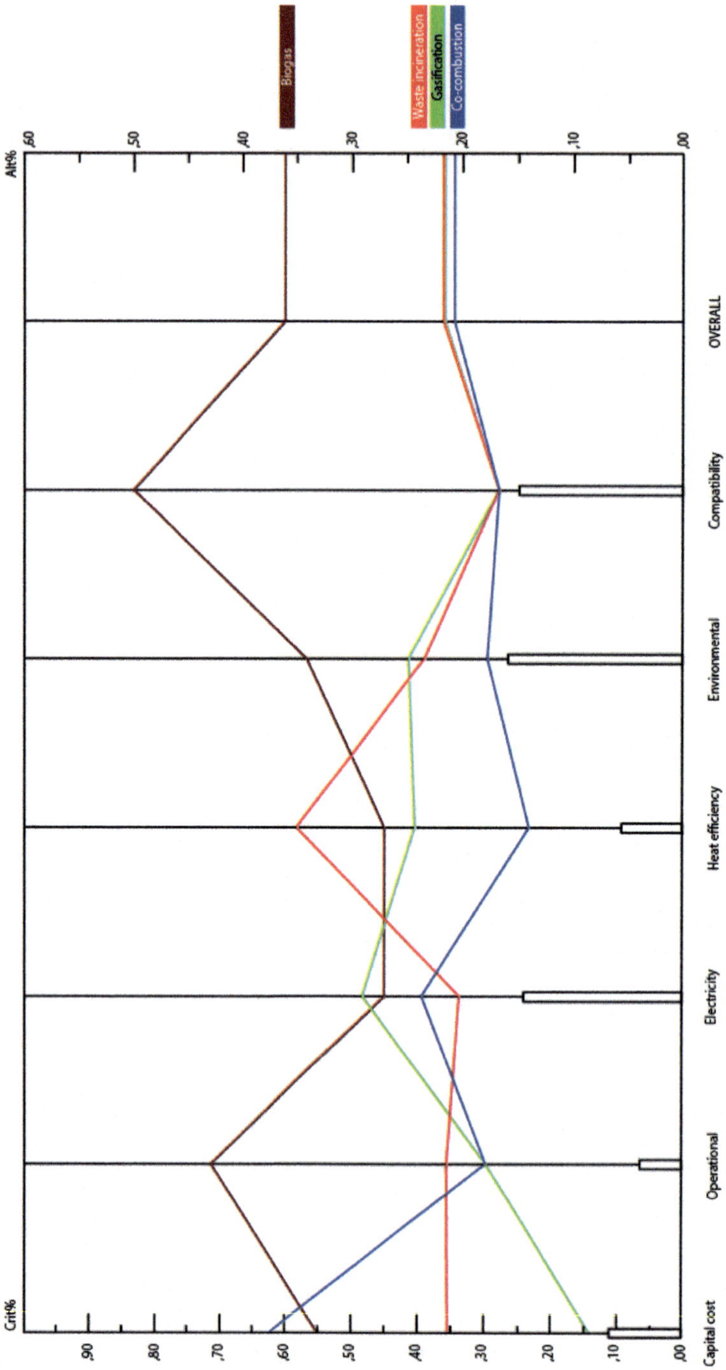

5.5 Conclusions

In a circular economy system, sustainability concerns waste prevention in the production and consumption phases, but sustainability supports recycling and, when appropriate, energy recovery in the waste management phase.[58] Initiatives to move towards a circular economy require adopting a holistic approach and complex systems thinking, focusing more on eco-effectiveness than on eco-efficiency.[59] CE requires considering not only solutions based on the life cycle of a process, but also the interaction between such a process and the environment and the economy in which it is embedded. Therefore, the assessment of the circularity of a technology or a management option requires to evaluate them referring to the complex socio-economic system with which they interact.

In order to contribute to the ongoing discussion on the feasibility of WtE technologies in the light of circular economy objectives, we performed an AHP assessment by considering four different technologies – co-combustion in coal-fired power plant; waste incineration with energy recovery; gasification; biogas production from anaerobic digestion – and three European countries: Denmark, the United Kingdom and Italy. In detail, we assessed such WtE technologies in three different scenarios, using six sub-criteria belonging to economic (i.e. capital costs and operational costs), circularity (i.e. electricity efficiency and heat efficiency) and environmental performance (environmental costs and compatibility with MSW). These criteria should offer a measure of the circularity of the selected WtE technologies, by taking into account economic indicators, energy recovery efficiency, environmental impacts and suitability with respect to waste streams and the separate collection system. The result shows that experts involved in the study have considered environmental criteria as the most important; meanwhile, heat efficiency and operational costs have been considered less important.

Moreover, this study highlights that in each country considered the most feasible WtE technology is different.

- gasification appears as the most feasible technology for energy generation from MSW in Denmark;
- in the United Kingdom the four technologies for energy generation from MSW scored almost at the same level (waste incineration with energy recovery is the most feasible solution followed by gasification and biogas production from anaerobic digestion);
- in Italy biogas production from anaerobic digestion was the most feasible solution (this result is mainly explained by the high compatibility of the technology with the high organic content of MSW).

---

58 A. Geldron, *'Économie circulaire: notions. Angers, Agence de l'Environnement et de la Maîtrise del'Energie'*, ADEME, 2013.
See also CIRAIG, *supra* note 4.

59 *Id.*

Several European reports and statistics[60] demonstrate that in a majority of Member States, energy and other resources contained in waste are being lost through improper management. Data show that the landfill rate is still too high and more efforts must be undertaken in order to meet the recycling targets and extract energy from non-recyclable waste in an efficient manner. Both energy and material recovered from non-recyclable waste fit in with resource circularity and will contribute to avoid landfill emissions thus supporting climate challenges. Furthermore, energy recovery from non-recyclable waste will help in the (at least partial) substitution of fossil fuels with renewable sources and therefore will contribute to energy security in the EU.[61]

AHP represents one of several methodologies that can be put in place in order to increase our knowledge about the feasibility of WtE technologies in a circular economy perspective, and will support technicians and decision-makers to fully exploit recovery of energy from non-recyclable waste. Similar studies, based on multi-criteria analysis, will contribute to compare the waste-to-energy situation in countries and to bring further clarity about the role of energy recovery in the EU Waste Hierarchy and the achievement of circular economy objectives.

With the aim to overcome the most important limitation of the present study, future research should consider social criteria (e.g. job creation) that are fundamental in a circular economy perspective. Basing future research on primary data will also help to better understand the underlying dynamics that cannot be fully captured by literature data.

---

60 *See* EUROSTAT, Treatment of waste by waste category, hazardousness and waste operations [env_wastrt], available from <http://tinyurl.com/qj7afre> (accessed 07-10-2016).
See also European Commission, Commission Staff Working Document – Impact Assessment Accompanying the document Proposal for a Directive of the European Parliament and of the Council amending Directives 2008/98/EC on waste, 94/62/EC on packaging and packaging waste, 1999/31/EC on the landfill of waste, 2000/53/EC on end-of-life vehicles, 2006/66/EC on batteries and accumulators and waste batteries and accumulators, and 2012/19/EU on waste electrical and electronic equipment, SWD(2014) 207 final, Brussels, 2014.
See also European Commission, Commission Staff Working Document – Ex-post evaluation of Five Waste Stream Directives Accompanying the document Proposal for a Directive of the European Parliament and of the Council reviewing the targets in Directives 2008/98/EC on waste, 94/62/EC on packaging and packaging waste, and 1999/31/EC on the landfill of waste, amending Directives 2000/53/EC on end-of-life vehicles, 2006/66/EC on batteries and accumulators and waste batteries and accumulators, and 2012/19/EC on waste electrical and electronic equipment, SWD(2014) 209 final, Brussels, 2014.

61 European Commission, *supra* note 27.

6 WtE and the Circular Economy: Environment and Energy

Thomas J. de Römph

6.1 Introduction

Over the past decades, trends such as globalization, growing world population, unsustainable manufacturing and use patterns, unsustainable energy production and use, and unsustainable waste management practices have been fostering human-induced pressures on natural resources, human health and the environment.[1] These pressures negatively affect not only the biosphere, which includes problems such as biodiversity loss, climate change and pollution of the hydrocycle, but also the non-biosphere. This includes issues such as competition for land and fresh water, the increasing demand for resources and the unpredictable and unsustainable supply of energy. The biosphere and non-biosphere cannot, of course, be seen separately from one another.

While being considered a problem in the past, it is now believed that waste can also offer solutions to some of these trends and pressures, in particular the ones concerning the energy system and the materials system. Resources are, after all, used both for energy supply and for materials intended for the technical sphere – *also* when they are contained in waste. In simple terms, using waste as a resource to generate energy would save fossil fuels such as coal, gas and oils used in conventional power plants. It would contribute to the security of energy supply, help to reduce the Union's dependency on fossil fuels (imports) and help us to get rid of unwanted waste and divert waste from landfills. Additionally, waste can also be used to replace virgin resources intended for the production of other materials, the manufacturing of products, the construction of buildings. The advantages of using waste as opposed to virgin resources are in fact quite similar to the advantages for the energy system, as it would contribute to the security of resources, help to reduce the dependency on virgin resource imports and help to divert waste from landfills. Besides, the emission of CO_2 and other polluting (greenhouse) gasses into the atmosphere is lowered in both cases. Both approaches are therefore generally considered as tools in the battle against climate change.

---

1 *See, e.g.* the reports of the Organisation for Economic Co-operation and Development (OECD) such as: OECD, 'Material Resources, Productivity and the Environment', *OECD Publishing*, 2015; and OECD, 'Sustainable Materials Management. Making Better Use of Resources', *OECD Publishing*, 2012.

The practice of using waste to produce energy is contained in the concept of "Waste-to-Energy" (WtE). Technically speaking, WtE covers all sorts of methods; some of which are addressed in this book. Without going into details of the technologies, it is sufficient to say that the WtE processes such as landfill gas production, incineration, thermal gasification and anaerobic digestion can,[2] depending on the technology and waste stream used of course, result in energy in the form of electricity, heat, gas and/or other (transport) fuels.

The European Union (EU, Union) has not been idle since these trends and pressures have deepened. Particularly in the last 15-20 years, the EU has launched a variety of policies that somehow deal with the triangular nature of material–waste–energy. Each time, these policies have been initiated in different policy contexts, highlighting different areas of concern and providing various, context-specific solutions. The policies for waste and for energy have been most influential in that respect. This two-pronged approach for WtE activities has evidently also contributed to the different viewpoints taken on resource use in EU environmental law (which includes EU waste law) on the one hand and EU energy law on the other hand.

In 2015, a new policy framework was launched by the European Commission (Commission): the Circular Economy Package (CE Package, Package).[3] It is significant for the Union's view on WtE practices, because it aims at the transformation of the economy of the EU into a so-called "Circular Economy" by 2050, "where nothing is wasted and where natural resources are managed sustainably."[4] Based on this very general objective, the Circular Economy transition, which has been given substance in the CE Package through the proposals of numerous policy and legal actions, may have the potential to influence the two frameworks for WtE, i.e. the one for waste and the one for energy. Establishing the right regulatory framework for enabling the transition to a more Circular Economy in the European Union is one of the crucial challenges faced by the EU.[5]

This contribution sheds light on what is considered to be the *right regulatory framework* for the Circular Economy transition and what impact this projected framework would have on WtE practices.[6] With this in mind, the existing regulatory framework

---

2 These four examples of WtE processes, which are the most common, are based on the life cycle assessment of municipal solid waste, *see*: A.U. Zaman, 'Life Cycle Environmental Assessment of Municipal Solid Waste to Energy Technologies', *Global Journal of Environmental Research*, 3(3), 2009, pp. 155-163. See also the report of the Joint Research Centre (JRC) for five (similar) main categories of WtE processes (combustion plants, waste incineration plants, cement and lime kilns, anaerobic digestion plants and others) and for the techniques for these categories: JRC: H. Saveyn, et al., 'Towards a Better Exploitation of the Technical Potential of Waste-to-energy', European Union, 2016.

3 *See*, for the main policy document: European Commission, *Closing the Loop – An EU Action Plan for the Circular Economy*, COM(2015) 614.

4 European Commission (2013). Decision 1386/2013 on a General Union Environment Action Programme to 2020 'Living well, within the limits of our planet', OJ L 354/171, Ann., 1.

5 European Commission, *supra* note 3, p. 2.

6 This contribution will not highlight one or more WtE technologies – it is technology neutral.

for waste is briefly explained. This provides an understanding of how WtE practices are currently regulated under EU waste law. Next, the Circular Economy Package is explained in order to present the proposals for changing the framework. It includes an explanation of the policy roots of the CE Package. Finally, a concluding discussion is provided.

6.2 EU Waste Legislation – Key to WtE Practices under EU Environmental Law

EU waste legislation consists of five pillars.[7] The first pillar comprises the framework legislation. This contribution predominantly focuses on the horizontally applicable Waste Framework Directive (WFD),[8] as it lays down definitions and instruments that are highly important for WtE activities. The second pillar concerns legislation on particular waste management operations. In this respect, both the Landfill Directive and the former Waste Incineration Directive are touched upon. The third pillar consists of legislation on specific waste streams. The Batteries Directive,[9] the Directive on End-of-Life Vehicles (ELV),[10] the Waste of Electrical and Electronic Equipment (WEEE) Directive[11] and the Packaging Directive[12] also enjoy a modest role in this contribution.

6.2.1 Waste Framework Directive

The Waste Framework Directive is most relevant for WtE practices. It lays down measures to protect the environment and human health by preventing or reducing the adverse impacts of the generation and management of waste, and by reducing the overall impacts of resource use and improving the efficiency of such use (Article 1 in conjunction with Article 13 WFD). Amongst others, it does so by setting the elementary concepts and definitions related to waste management, such as the definitions for waste, recycling and energy recovery. The Directive also provides for certain instruments that are relevant for WtE practices, such as the waste hierarchy and the waste management targets.

---

7 This division is based on the website of the European Commission on EU waste legislation: <http://ec.europa.eu/environment/waste/legislation/>, consulted on 25 October 2016. The final two pillars are not discussed in this contribution (i.e. reporting and questionnaire legislation, and useful links to other directly or indirectly related legislation).

8 Directive 2008/98 of 19 November 2008 on waste and repealing certain Directives, OJ L 312/3.

9 Directive 2006/66 of 6 September 2006 on Batteries and Accumulators and Waste Batteries and Accumulators and Repealing Directive 91/157/EEC, OJ L 266/1.

10 Directive 2000/53 of 18 September 2000 on End-of Life Vehicles, OJ L 269/34.

11 Directive 2012/19 f 4 July 2012 on Waste Electrical and Electronic Equipment (WEEE), OJ L 197/38.

12 Directive 94/62 of 20 December 1994 on Packaging and Packaging Waste, OJ L 365/10.

The waste hierarchy and some associated definitions and waste management targets are explained.

6.2.1.1 Waste Hierarchy

Member States are encouraged to adopt waste management measures that deliver the best overall environmental outcome. This is done by the employment of a life cycle approach, amongst others through the waste hierarchy (Article 4(1) WFD). The hierarchy had originally been developed to divert waste from landfilling (since 1975), but has along the way been adjusted considering other objectives, too. It is now the corner stone of EU waste legislation. In the 2008 version of the Waste Framework Directive, the hierarchy offers policy makers and legislators in Member States a more enhanced priority order for waste management options. The five-step waste hierarchy has the following options:

a. prevention
b. preparing for reuse
c. recycling
d. other recovery, e.g. energy recovery
e. disposal

Because the hierarchy is based on generalization, the order can be customized for a specific waste stream but only if this is justified by "life cycle thinking" on the overall impacts of the generation and management of such waste (Article 4(2) WFD).[13] This opportunity to deviate rests on the idea that the hierarchy cannot *always* provide good guidance. Accordingly, as long as national authorities follow the waste hierarchy in their policy and regulation, a proper application of life cycle thinking is assumed.[14] And because the concept of life cycle thinking is not specified in the Waste Framework Directive or in any other EU waste measure, the interpretation and application of the waste hierarchy and life cycle thinking by the Member States is quite flexible on condition that the choices made are explained thoroughly.[15]

---

13 *See* Section 6.3.3 for a general explanation of this concept.

14 European Commission, *Guidance on the interpretation of key provisions of Directive 2008/98/EC on waste*, 2012, p. 49, see <http://ec.europa.eu/environment/waste/framework/pdf/guidance_doc.pdf> (consulted 15 November 2016).

15 *See, e.g.* J. Wante, 'A European Legal Framework for Enhanced Waste Management', *1st Int. Symposium on Enhanced Landfill Mining*, 2010, p. 5; and C. Dalhammar, 'The Application of "Life Cycle Thinking" in European Environmental Law: Theory and Practice', *Journal for European Environmental & Planning Law*, 12, 2015, pp. 106 and 114-118. *See also* for the particular application of life cycle thinking in the waste hierarchy, for example through life cycle assessments (LCAs), and for the challenges that go along with basing national polices and legislations on these LCAs and other assessments and life cycle thinking in general: D. Lazarevic, N. Buclet and Nils Brandt, "The Application of Life Cycle Thinking in the Context of European Waste Policy', *Journal of Cleaner Production*, 29-30, 2012, pp. 199-207.

6.2.1.2 WtE and the Waste Hierarchy

The concept of WtE is neither referred to nor explained in either the Waste Framework Directive or any other legislation on a particular waste management operation or on a specific waste stream. Questions arise, therefore, on how to define "WtE" and where to place WtE practices in the waste hierarchy.[16] The definitions provided in the Waste Framework Directive are crucial for answering these questions, in particular the one for recovery. According to Article 3(15) WFD, recovery means:

> any operation the principal result is waste serving a useful purpose by replacing other materials or waste being prepared to fulfil that function, or waste being prepared to fulfil that function, in the plant or in the wider economy.

This definition lays the foundation for the three successive recovery categories (b)-(d) in the hierarchy: "preparing for reuse", "recycling" and "other recovery". According to Article 3(17) WFD, recycling means:

> any recovery operation by which waste materials are reprocessed into *products, materials or substances* whether for the original or other purposes. It includes the reprocessing of organic material *but does not include energy recovery and the reprocessing into materials that are to be used as fuels or for backfilling operations.*[17]

A clear Waste-to-Material (WtM) is aimed for in recycling processes, as the primary objective of recycling is to add another cycle to the terminating life cycle of a material, which is done by aiming at new substances, products or materials.[18] It follows, unsurprisingly, that the definition explicitly states that recycling *excludes* energy recovery and the production of waste-derived fuels. Together with backfilling, these two recovery activities fall under the final category ("other recovery") for which no explicit definition exists in Article 3 WFD. Accordingly, one has to look at the general meaning of recovery, as

16 These questions have also been addressed in: T. de Römph and L. Reins, 'Waste-to-Energy and the Circular Economy – Connecting the Dots', *Oil, Gas, Energy Law Intelligence*, 14(3), 2016, pp. 5-8.

17 Emphasis added.

18 An exception to this rule is made in Article 2(6) Decision 2011/753 of 18 November 2011 establishing rules and calculation methods for verifying compliance with the targets set in Article 11(2) of Directive 2008/98/EC of the European Parliament and of the Council, OJ L 310/11. The provision states that for the purposes of verifying compliance with the targets set in Article 11(2) of the Waste Framework Directive (*see* Section 6.2.1.3) and where the target calculation as laid down in Decision 2011/753 is applied to the *aerobic or anaerobic treatment of biodegradable waste*, the input to the aerobic or anaerobic treatment may also be counted as recycled but only where that treatment generates *compost or digestate* which, following any further necessary reprocessing, is used as a recycled product, material or substance for land treatment resulting in benefit to agriculture or ecological improvement. Or in other words, when aerobic or anaerobic treatment of biodegradable waste generates compost or digestate which is used as fertilizer.

quoted above, minus the two recovery categories which are higher ranked in the hierarchy.[19] It should be noted, moreover, that the operations falling within the residual category "other recovery" are on equal footing: there is in principle no hierarchy between practices such as backfilling (which aims at WtM instead of WtE), waste-derived fuels or energy recovery on condition that the relevant waste stream is unfit for the preparation for reuse and recycling.

Because it is generally understood that WtE means energy is generated from waste in the form of electricity, heat, gas and/or *other fuels*,[20] it is evident that waste-derived fuels fall under this broad description, too. This type of waste-derived fuel is, however, often denied as a WtE operation: energy recovery alone is almost automatically associated with WtE practices.

While "energy recovery" has been expressly mentioned in the waste hierarchy as a subcategory of other recovery forms in Article 4(1) WFD, the Waste Framework Directive does not include a definition of the concept. Annex II WFD contains, however, a non-exhaustive list of recovery operations. This list provides certain energy efficiency thresholds for municipal solid waste incineration plants.[21] These thresholds are important because they determine whether WtE operations treating municipal waste either enjoy a recovery status or a disposal status.[22] This difference has consequences for the applicability of certain rules on both at EU level and at Member States level; for example, concerning the shipment of waste (see the Waste Shipment Regulation)[23] or national permits.[24]

---

19 The first recovery category is "preparing for reuse" and is insignificant in the discussion on WtE, because it aims at *returning* the waste as non-waste products to the use stage of a material's life-cycle. Article 3(16) defines it as: "cleaning or repairing recovery operations, by which products or components of products that have become waste are prepared so that they can be re-used without any other pre-processing." This recovery category therefore does not aim at the generation of materials or energy from waste.

20 *See* the introduction to this contribution.

21 The R1 formula (*see* below) is created for municipal solid waste, because this waste stream is most commonly used for fuels due to their availability and heterogeneity. Other, more homogeneous streams tend to be used more in WtM recovery operations. Nevertheless, these other waste streams can be used in WtE production respecting the waste hierarchy. T. Turunen, 'Efficient Waste-to-Energy Production in the European Union – The R1 Efficiency Standard and the Climate Correction Factor', *Oil, Gas, Energy Law Intelligence*, 14(3), 2016, p. 6. The fact that there is only a formula for municipal solid waste has furthermore been criticized. *See:* E. Stengler, 'Waste-to-Energy in the Circular Economy and the Energy Union', *Oil, Gas, Energy Law Intelligence*, 14(3), 2016, pp. 1-13. Stengler points out that the R1 formula does not apply to commercial and industrial waste even if these waste streams are similar to municipal solid waste and are not suitable to prepare for reuse and recycling. She argues that this would hinder the Circular Economy transition, because these waste streams would be more easily consigned to a landfill than to an energy recovery plant, which not corresponds to the waste hierarchy.

22 The Court of Justice has decided that a waste treatment operation can either have a recovery *or* a disposal status. Case C-6/00 *Abfall Service AG v. Bundesminister fur Umwelt, Jugend and Familie* [2002] ECR I-1961, para. 62.

23 Regulation 1013/2006 of 14 June 2006 on Shipments of Waste, OJ L 190/1. *See e.g.* Article 3(1)(a), (2) and (5).

24 *See also* for the consequences of the recovery or disposal status: Turunen, *supra* note 21, p. 3.

Just like "other recovery" is a residual category of "recovery", "disposal" is a residual category of "waste treatment".[25] According to Article 3(19) WFD, disposal means any operation that is *not* recovery even where the operation has as a secondary consequence the reclamation of substances *or energy*.[26] The set of energy efficiency thresholds mentioned above is based on the so-called "R1 formula".[27] The formula boils down to the idea that for energy recovery the principal result is the generation of a considerable amount of energy, whereas disposal incineration facilities, for example, generate low energy volumes.[28] For other disposal activities there are *no* energy efficiency thresholds established.

Considering the above, it is clear that WtE operations can be classified either as recovery *or* as disposal under the Waste Framework Directive.[29] This distinction depends on whether they fulfil the requirements of the definition of recovery and, if any, they meet certain energy thresholds. All in all, it can also be concluded that while WtE activities can be ranked differently in the waste hierarchy, WtM activities generally prevail over WtE activities. This general rule is not easily overridden by life cycle thinking, because the hierarchy may only be altered for *specific* waste streams. This issue was raised in the drafting process of the German federal Closed Substance and Cycle and Waste Management Act.[30] Article 8(3) of the proposed Act ranked material recovery (i.e. preparing for reuse, recycling and WtM other recovery options) and energy recovery equally, provided that the waste stream had a calorific value of at least 11,000 kilojoules per kilogram. It is for precisely this reason why the European Commission raised its concern: that the deviation from the waste hierarchy was not limited to a specific waste stream but, instead, was generalized to all waste streams.[31] It stated that the Act:

25 According to Article 3(14), waste treatment means "recovery or disposal operations, including preparation prior to recovery or disposal."

26 *See also* Ann. I WFD for examples of disposal operations.

27 *See* the non-legally binding guidance: European Commission, *Guidelines on the interpretation of the R1 energy efficiency formula for incineration facilities dedicated to the processing of MSW* (European Commission, 2011). *See also* the Cases, *e.g.*: C-6/00 *Abfall Service AG v. Bundesminister fur Umwelt, Jugend and Familie* [2002] ECR I-1961; and C-228/00 *Commission v. Germany* [2003] ECR I-1439.

28 *See also* Article 23(4) WFD in which it is stated that it shall be a condition of any permit covering incineration or co-incineration with *energy recovery* that the recovery of energy takes place with a high level of energy efficiency. The energy recovery sector is thereby the first and only sector for which an efficiency criterion has been introduced in the Directive and the quality of the energy recovery has improved as a result: *see* CEWEP, 'Energising Waste – A Win-win Situation', available at <www.cewep.eu/information/policy/m_1473>, p. 2 (consulted 15 November 2016).

29 But *see supra* note 18.

30 *See* the 'Gesetz zur Neuordnung des Kreislaufwirtschafts- und Abfallrechts Kreislaufwirtschaftsgesetz, KrWG', Federal Law Gazette ('Bundesgesetzblatt'), published 29 February 2012.

31 Dalhammar, *supra* note 15, pp. 16-18. Dalhammar provides for examples in which a lower national standard would be aimed after. For example, he points out the situation where a Member State may have few facilities for recycling, which means that recycling requires transport of waste to recycling facilities faraway. If there are incineration plants near the place of waste production, a life cycle assessment could show that incineration is the preferable option from an environmental point of view. According to him, this is however a *static way* to view the issue, because if more recycling facilities are set up and more waste go to these plants,

> could therefore lead to a weakening of the priority given for preparation for re-use and recycling which is not in line with the waste hierarchy and the objectives of Directive 2008/98/EC and has therefore invited the German authorities to revise Article 8 of the notified draft act in order to better reflect the concept of waste hierarchy [...].[32]

Article 4(2) WFD, in which life cycle thinking is provided as a way to deviate from the waste hierarchy, should thus be applied with caution, only in specific cases, and not become an excuse to have a generally lower ambition level in national waste policy.[33]

6.2.1.3 WtE and Waste Management Targets

The Waste Framework Directive and other EU waste measures contain various binding waste management targets. These targets are established to move towards a so-called "recycling society"[34] and to set the waste hierarchy in motion.

As for the Waste Framework Directive, Article 11(2) sets targets for waste that must be prepared for reuse and recycled to increase the level of resource efficiency in the EU. Member States must take necessary measures to achieve the following targets by 2020. 50% by weight of household waste (such as at least paper, metal, plastic and glass) and possibly waste from other origins if similar to household waste, must be prepared for reuse and recycled. 70% by weight of non-hazardous construction and demolition (C and D) waste (excluding naturally occurring material defined in Category 17 05 04 in the list of waste) must be prepared for reuse, recycled and recovered in another way for its materials, such as backfilling operations using waste to substitute other materials.

While there are general targets for preparing for reuse and recycling of specific waste streams, there are *no* targets in the Waste Framework Directive that explicitly or impli-

---

recycling could become an environmentally and economic more viable option than incineration over time. The waste hierarchy therefore requires a *long term vision*.

32 Commissioner Potocnik provided a statement on behalf of the Commission and on request from a member of the European Parliament, where he expressed concerns about the draft Act. *See*: Parliamentary questions of 4 August 2011, <www.europarl.europa.eu/sides/getAllAnswers.do?reference=E-2011-005520&language=SV> (consulted on 13 November 2016).

33 Dalhammar, *supra* note 15, p. 18. Either way, Germany did not change the wording of Article 8(3) in the final text of the Closed Substance and Cycle and Waste Management Act. *See also e.g.* a letter of several German environmental NGOs, such as Nature And Biodiversity Conservation Union ('NABU') of 27 April 2012 on the concerns about the implementation of the waste hierarchy in the Act: <www.nabu.de/imperia/md/content/nabude/konsumressourcenmuell/120427_greenngo_letter_dgenvi_wfd_germany.pdf> (consulted on 13 November 2016).

34 Recitals (28) and (41) and Article 11 WFD.

citly aim at WtE.[35] The waste management targets are deeply rooted in the concepts of preparing for reuse, recycling and other WtM recovery routes.[36]

6.2.2 *Legislation on Particular Waste Management Operations*

Both the Landfill Directive and the Waste Incineration Directive (including the Industrial Emissions Directive) are briefly explained in this section.

6.2.2.1 Landfill Directive

The Landfill Directive (LFD)[37] applies to all landfills.[38] The Waste Framework Directive does not give a definition of landfilling, only a general definition of disposal is provided.[39] Annex I WFD, however, sets out a list of disposal operations. This list includes certain landfill operations, such as an ordinary landfill (entry D1) and a specially engineered landfill (D5).

The overall aim of the Directive is "to prevent or reduce as far as possible negative effects on the environment from the landfilling of waste [...] during the whole life cycle of the landfill" through stringent operational and technical requirements for landfill sites.[40] For example, it lays down certain strict requirements for the closure of and aftercare for landfill facilities.[41] During the life time of a landfill site, other requirements apply. Besides posing several controlling and monitoring mechanisms, the Landfill Directive makes considerable efforts to reduce biodegradable waste going to landfills (by means of in

---

35 And hence, the fact that there are no waste management targets that distinguish between WtE recovery and disposal operations and non-WtE disposal operations is open to question. T. Turunen and G. Van Calster, 'Burning Construction and Demolition Waste: An Assessment of the European Waste Hierarchy and Recycling Targets', *Oil, Gas, Energy Law Intelligence*, 14(3), 2016, p. 8.

36 *Id.*, at p. 5. In view of this, Turunen and Van Calster stress that these targets do not include any waste prevention targets, even though prevention targets (which also aim at WtM) would fit the waste hierarchy. The main reason why prevention targets are not established is the difficulty of calculating how much waste is prevented from being produced.

37 Directive 1999/31 of 26April 1999 on the landfill of waste, 16 July 1999, OJ L 182/1.

38 Article 3(1) LFD.

39 Because the Waste Framework Directive lacks to do so, the Landfill Directive defines landfilling instead. According to Article 2(g) of the Directive, a landfill is:
"a waste disposal site for the deposit of the waste onto or into land (i.e. underground), including:
- internal waste disposal sites (i.e. landfill where a producer of waste is carrying out its own waste disposal at the place of production), and
- a permanent site (i.e. more than one year) which is used for temporary storage of waste,

but excluding:
- facilities where waste is unloaded in order to permit its preparation for further transport for recovery, treatment or disposal elsewhere, and
- storage of waste prior to recovery or treatment for a period less than three years as a general rule, or storage of waste prior to disposal for a period less than one year."

40 Article 1(1) LFD.

41 Article 13 LFD.

particular recycling, composting, other forms of material recovery, biogas production and energy recovery) and to ban certain types of waste.[42]

The possibilities to employ and develop WtE activities – amongst others via the capturing and processing of gasses – are clearly not shunned or overlooked in the Landfill Directive. Despite these opportunities for WtE, it should be noted that landfilling always remains a disposal activity, ranked last in the waste hierarchy, even if energy is generated from the landfilled waste.[43]

6.2.2.2 From the Waste Incineration Directive to the Industrial Emissions Directive

The aim of the Waste Incineration Directive[44] was to prevent or to limit as far as practicable negative effects on the environment, in particular pollution by emissions into air, soil, surface water and groundwater, and the resulting risks to human health, *from the incineration and co-incineration of waste* (Art. 1 WID).[45] Stricter operational conditions and technical requirements were set than before: e.g. emission limit values for waste incineration and co-incineration plants. Both new and existing WtE plants, *both intended for recovery as well as disposal*, had to conform to these new limits, also when they dealt with domestic waste. The limits in the Waste Incineration Directive had to be made subject to a separate permit.

As from 6 January 2011, the Industrial Emissions Directive (IED)[46] integrated the Waste Incineration Directive and five other directives, e.g. the Integrated Pollution Prevention and Control Directive. Consequently, there is an increase of industrial installations subject to the Directive. Its aim is to achieve significant benefits for the environment and for human health by reducing harmful industrial emissions across the EU, in partic-

---

42 Article 12 and Ann. III and Article 5(1), (2) and (3) LFD, respectively. The types of waste not accepted are for example hospital and other clinical waste which is infectious, liquid waste and tyres (with certain exceptions).

43 *See e.g.* European Commission, *supra* note 14, p. 35.

44 Directive 2000/76 of 4 December 2000 on the Incineration of Waste, [2000] OJ L 332/91. The Directive replaced three Directive, *e.g.* on the incineration of hazardous waste.

45 According to Article 3(4) and (5), an incineration plants means "any stationary or mobile technical unit and equipment dedicated to the thermal treatment of wastes *with or without recovery* of the combustion heat generated. This *includes* the incineration by oxidation of waste as well as other thermal treatment processes such as pyrolysis, gasification or plasma processes in so far as the substances resulting from the treatment are subsequently incinerated" (emphasis added), and co-incineration means "any stationary or mobile plant whose *main purpose is the generation of energy or production of material products* and: – which uses wastes as a regular or additional fuel; or – in which waste is thermally treated for the purpose of disposal. If co-incineration takes place in such a way that *the main purpose of the plant is* not the generation of energy or production of material products but *rather the thermal treatment of waste, the plant shall be regarded as an incineration plant* [...]" (emphasis added). *See also* Case C-317/07 *Lahti Energia Oy* [2010] ECR I-1429. The Directive contained specific provisions for co-incineration plants.

46 Directive 2010/75 of 24 November 2010 on Industrial Emissions (Integrated Pollution Prevention and Control), OJ L 334/17.

ular through better application of Best Available Techniques (BATs) (Art. 1 in conjunction with Art. 11 (b) IED). Operators of industrial installations' operating activities covered by a long list in Annex I of the IED are required to obtain an integrated permit from the authorities in the EU countries to prevent and control the pollution arising from their activities. Under the Industrial Emissions Directive the definition of "incineration plant" and "co-incineration plant" remain basically the same (Article 3(40) and (41) IED). Importantly, the Directive puts forward several general principles governing the basic obligations of the operators: according to Article 11(d) and (e) IED, Member States shall take the necessary measures to ensure that these plants are operated in accordance with the waste hierarchy.

6.2.3 *Waste Management Targets in Specific Waste Stream Measures*

Not only does the Waste Framework Directive contain waste management targets, targets are also set in measures regulating specific waste streams. Directive 2000/53 on end-of-life vehicles, for example, aims at the recovery of ELVs so as to reduce the disposal of waste and to improve the environmental performance of all of the economic operators involved in the life cycle of vehicles (Article 1 ELV Directive), amongst others by setting targets for preparing for reuse, recycling and other recovery forms. Article 7(2)(b) of the Directive states, for example, that no later than 1 January 2015, recovery (so including WtE practices) of all ELVs shall be increased to a minimum of 95% by an average weight per vehicle and year. Within the same time limit, preparation for reuse and recycling shall be increased to a minimum of 85% by an average weight per vehicle and year.[47]

The Batteries Directive regulates the use of hazardous substances in batteries and lays down rules for the collection, treatment, recycling and disposal of waste batteries and to promote a high level of collection and recycling of these batteries (Article 1). Regarding the latter set of rules, the Batteries Directive sets an overall collection target of 45% by 26 September 2016 (Article 10(2)(b)) and provides for the commitment that recycling processes shall meet certain recycling efficiencies no later than 26 September 2011 (Article 12 (4) in conjunction with Annex III, Part B).

The WEEE Directive regulates the management of WEEE. In conjunction with Annex V of the Directive, Article 11 (1) sets out a number of minimum targets that producers must meet. With no exception, for each predefined WEEE stream there are two targets: one for recovery in general and one specifically for either recycling alone or for preparing for reuse and recycling. Moreover, there are three time frames for these targets during

47 The word "reuse" is used incorrectly in the ELV Directive (Art. 2(6)), as it mismatches the definitions put forward in the Waste Framework Directive (Art. 3(13)). The ELV word for "reuse" matches, however, the WFD definition for preparing for reuse. For this reason, this contribution refers to "preparing for reuse" even though the text of the ELV Directive uses "reuse".

which the targets may be increased. For lighting equipment for example, 70% shall be recovered and 50% shall be recycled between August 2012 and August 2015. Between August 2015 and August 2018, 75% shall be recovered and 55% shall be prepared for reuse and recycling and from August 2018, 75% shall be recovered and 55% shall be prepared for reuse and recycling.

The Packaging Directive lays down measures aimed at putting the waste hierarchy into practice for packaging waste (Article 1(2)). To this end, the Directive contains several targets, for example on the prevention of packaging waste (Article 4(1a)(a) and (b)). For the recovery of such waste, Article 6 lays down targets for two time frames (until 2001 and until 2008), and for each period one target for recovery in general and one for recycling alone.[48] The recycling target for the final period is furthermore divided into several sub-targets for specific waste streams. For example, no later than December 2008 the minimum recycling targets for glass contained in packaging waste is 60% by weight.[49]

6.3 The Transition towards a Circular Economy in EU Law

After having described the state-of-the-art for WtE activities in EU waste law in Section 6.2, this section sheds light on the projected framework under the Circular Economy Package. This requires a five-layered approach. First, by way of introduction, the most important policies of the CE Package and therefore also for the Circular Economy transition are explained. They are significant in the sense that they have coined the concept in formal EU policy and legislation and that they lay the foundation for the Circular Economy Package by introducing fundamental concepts, which, as a consequence, simultaneously give an indication on the Commission's position on WtE in the future. Second, a description of the Circular Economy concept under the CE Package is provided and its most important aspect is explained next. Finally, the changes to the current regulatory framework that are significant for WtE activities are set forth.

6.3.1 The Rise of the Circular Economy in EU Environmental Policy

The concept "Circular Economy" did not come out of the blue. Many environmental policies have inspired the development of the Circular Economy Package. Although they underpin the Package, these policies had not initially been developed to facilitate

48 For the first period, Article 6(1)(a) and (b) of the Packaging Directive state that no later than 2001 between 50% as a minimum and 65% as a maximum by weight of packaging waste will be recovered, and no later than the same date between 25% as a minimum and 45% as a maximum by weight of the totality of packaging materials contained in packaging waste will be recycled with a minimum of 15% by weight for each packaging material.

49 Article 6(1)(e)(i) Packaging Directive.

the transition towards a Circular Economy. Their origins can be traced back to different constituent aspects of the Circular Economy, and are based on various overlapping or interlinked grounds, involving different actors and covering a different mix of policy areas.[50] Over the years, some of these emerging policies have been quite successful in building the foundation for the paradigm shift, which is aimed at by the Circular Economy ideology. They thereby already influenced a whole host of legal measures in the EU, including the Waste Framework Directive. Three prominent policy frameworks are explained below in chronological order.

6.3.1.1 The Flagship Initiative for a Resource-Efficient Europe

The transition towards a Circular Economy was coined for the first time in the 2001 Flagship Initiative for a Resource-Efficient Europe,[51] which is a policy framework that has been created under the broader Europe 2020 Strategy.[52] This broader framework was set up by the Commission as a 10-year growth strategy in 2010 to emerge stronger from the economic and financial crisis in the EU. Not surprisingly, therefore, it wanted to create a *smart, sustainable and inclusive economy*. Europe 2020 aims at short-term goals (i.e. exit from the economic crisis) and the future-orientated goal of sustainability. The Commission proposes targets for five policy fields that should steer that process until 2020: employment; research and innovation; education; poverty; *climate change and energy*. To achieve these priority policy fields and to give further substance to Europe 2020, the Commission initiated seven Flagship Initiatives, one of which is the Resource-Efficiency Europe Initiative.[53]

The implementation of the Resource-Efficiency Initiative is the responsibility of the Directorate-General (DG) for the Environment (DG ENV) of the European Commission. Note that this was *neither* the former DG ENER *nor* the DG for Climate Action (CLI-

50 Not only is the Commission's view on the Circular Economy transition inspired by existing policy tracks in the EU, also some other international organizations (such as the OECD's policy regarding the concept of "Sustainable Materials Management", *see e.g.* the document in *supra* note 1) developed intimately related concepts and policies that influenced the CE Package. After all, the trends and pressures pointed out in the introduction of the contribution occur globally.

51 European Commission, *A resource-efficient Europe – Flagship initiative under the Europe 2020 Strategy*, COM(2011) 21; and European Commission, *Roadmap to a Resource Efficient Europe*, COM(2011) 571.

52 European Commission, *Europe 2020 – A strategy for smart, sustainable and inclusive growth*, COM(2010) 2020.

53 Other Flagship Initiatives concern amongst others the "Innovation Union", which tries to improve the framework conditions and access to finance for research and innovation (European Commission, *Europe 2020 Flagship Initiative Innovation Union*, COM(2010) 546) and the "An Agenda for New Skills and Jobs" initiative (European Commission, *An Agenda for new skills and jobs: A European contribution towards full employment*, COM(2010) 682). Besides the Resource-Efficient Europe Initiative these policy frameworks are also important for the transition towards a Circular Economy. They will however not be discussed in more detail here, because they have not explicitly mentioned the need to transform the economy into a Circular Economy.

MA),[54] which could also have been suitable DGs for action on resource-efficiency. Arguably, the fact that DG ENV took the responsibility for the implementation of the Flagship Initiative can be explained by the reshuffling of the responsibilities of the DGs when the second Barroso Commission took office in 2010. Before 2010, DG ENER used to take on the responsibility for energy and DG ENV for the environment *including* climate change. But as climate change moved to a DG expressly created to control harmful emissions at EU and international level (i.e. DG CLIMA), DG ENV had to search for other emblematic topics to take up its mandate: that is resource-efficiency.

At the most basic conceptual level, resource-efficiency means "doing more with less resources". The Resource-Efficiency Initiative is not exclusively restricted to the transition of the materials system, however. Despite the separated responsibilities for the environmental-orientated DGs at that time, the policy also aims at other issues, such as stimulating clean and efficient energy, combating climate change, sustainable fisheries, regional development and protecting biodiversity. "Resources" should therefore in principle be interpreted broadly in the context of the Initiative. Moreover, this interpretation can also be acquired from the more extensive definition of "resource-efficiency", which is:

> to help decouple economic growth from the use of resources, support the shift towards a low carbon economy, increase the use of renewable energy sources, modernise our transport sector and promote energy efficiency.[55]

Having said that, because the decoupling of economic growth from resource use is utilized distinctively from the energy system (i.e. by separately mentioning the promotion of energy-efficiency and the use of renewable energy sources), there are reasonable grounds to believe that "resource decoupling" particularly aims at making the *materials* system more sustainable.[56] This indicates that while the Initiative targets many environmental issues related to resource use, including the ones concerning energy supply and use, resource decoupling only relates to resources intended to fulfil a function in the materials system.

To achieve resource-efficiency in the European Union, the Flagship Initiative mentions amongst other actions the launch of:

---

54 In 2011, there were two distinctive DGs for action on climate change and for action on energy.

55 European Commission, *supra* note 52, p. 6. Other objectives put forward in the Resource-Efficient Europe Initiative are amongst others stimulating technological improvements and changing the behaviour of producers and final users. European Commission, *supra* note 51, p. 3. Indeed, the latter goal refers to the SCP Action Plan.

56 By the way, the Resource-Efficient Europe Initiative, of course, also aims at environmental impact decoupling from growth as a distinctive goal through the determination of medium and long-term sub-objectives and actions. European Commission, *supra* note 51, p. 5.

> [a] strategy to make the EU a "*circular economy*", based on a *recycling society* with the aim of reducing waste generation and using waste as a resource.[57]

What is more is that a transitional approach is adopted in the Initiative in the sense that the Commission foresees that the completion of the Resource-Efficiency Initiative, and thus including the quest to create a Circular Economy, is achieved in 2050. Meanwhile, incremental steps should be taken by 2020 to speed up the transformation. This approach is said to provide a clear path and ensure long-term predictability for businesses and investors to boost the competitiveness of the EU. Needless to say, it was to be expected that there would be a follow-up on the matter of the Circular Economy, as the Initiative was not quite clear on what a "Circular Economy" really means other than that this would be based on a "recycling society".

6.3.1.2 The 7th Environment Action Programme

Two years later than the Resource-Efficiency Initiative, the most recent Environment Action Programme (EAP),[58] the 7th EAP, was adopted by the name "Living well, within the limits of our planet".[59] Besides recalling the objectives laid down in the 6th EAP[60] and trying to address the shortcomings in the implementation of these objectives, the 7th EAP puts forward a renewed vision for the environmental policy of the EU. The main goal is formulated in the first paragraph of the Annex of the 7th EAP and reads as follows:

> In 2050, we live well, within the planet's ecological limits. Our prosperity and healthy environment stem from an innovative, *circular economy* where nothing is wasted and where natural resources are managed sustainably, and biodiversity is protected, valued and restored in ways that enhance our society's resilience. Our low-carbon growth has long been decoupled from resource use, setting the pace for a safe and sustainable global society.[61]

57 Emphasis added. *Id.* at p. 6.

58 On irregular but well-thought-out moments, for instance on the eve of some major international conference, the EU sets out priority objectives to be attained in EAPs. These EAPs have been adopted as legally binding decisions since the 6th EAP, based on Article 192(3) TFEU, which makes their content official sources of Union law.

59 Decision 1386/2013 of 20 November 2013 on a General Union Environment Action Programme to 2020 "Living well, within the limits of our planet", OJ L 354/171.

60 Decision 1600/2002 of 22 July 2002 laying down the Sixth Community Environment Action Programme, OJ L 242/1. *See also* European Commission, *Environment 2010: Our future, Our choice*, COM(2001) 31. The 6th EAP already covers a number of issues that are significant in the Circular Economy but which are not put forward in this contribution.

61 Emphasis added.

What makes the 7th EAP a milestone document is that it coins the "Circular Economy" for the first time a legally binding source of EU legislation. The Programme further pictures the Circular Economy for 2050.[62] Just as the Resource-Efficient Europe Initiative, which also pictures full achievement of resource-efficiency in 2050, the 7th EAP does not explain the meaning of the Circular Economy transition. A series of features may nevertheless be understood from the priority objectives that are provided in Article 2 of the 7th EAP (which are further clarified in its Annex). The first three priority objectives are most important. They are: (1) protecting, conserving and enhancing the Union's natural capital; (2) turning the Union into a resource-efficient, green and competitive low-carbon economy; and (3) safeguarding the Union's citizens from environment-related pressures and risks to health and well-being.[63]

Regarding the second objective, which neatly ties in with the Research-Efficient Europe Flagship Initiative, paragraph 29 Annex of the 7th EAP stresses that the EU needs to stimulate growth by deploying a more competitive economy that uses resources sustainably over the entire life cycle of a material. It would do so, if GHG emissions are reduced, competitiveness is enhanced through efficiency and eco-innovation, and energy and resources are secured, including through reduced overall resource use. In view of this, moreover, according to the EAP, it would particularly help if energy decoupling, environmental impact decoupling and resource decoupling are stimulated – indeed, the same distinction is put forward as in the Research-Efficient Initiative. On the whole, the 7th EAP does not specifically target the materials system. Issues such as biodiversity loss and water scarcity, climate change and inefficient energy supply and use, and the non-renewable use of energy resources should also to some extent be targeted in a Circular Economy (see, e.g. paragraphs 33 and 43(a) and (i) Annex 7th EAP). In that sense, the Circular Economy movement can be framed within a broader ambition of "sustainability", which is also adopted in the Resource-Efficiency Europe Initiative.

That being said, however, paragraph 40 indicates that waste should be turned back into a resource, as it can be used again in the economy; the aim to move towards a recycling society is clearly prevalent. Notably, the paragraph explicitly mentions the term "Circular Economy". In the light of this aim, the 7th EAP states that amongst others additional efforts are required: to reduce waste generation in absolute terms and per capita; to limit energy recovery from waste to non-recyclable materials; to phase out landfilling of recoverable waste; to ensure high-quality recycling where the use of recycled material does not lead to overall adverse environmental or human health impacts; and to develop markets for these recycled materials. All in all, the waste hierarchy is properly

62 The 7th EAP further highlights that the achievements of the Programme's objectives for 2020 should be in line with the Circular Economy vision for 2050, as it would provide a stable environment for sustainable growth (*see* Recital (8) 7th Environment Action Programme). It thereby foresees that several incremental steps need to be taken in order for the existing economy to become a Circular Economy.

63 Article 2(1)(a)-(c) 7th EAP.

applied while putting extra emphasis on high-quality recycling. Besides the call for better "circular" *waste* management, the 7th EAP also stresses in paragraph 35 of its Annex the need for more environmentally sustainable *production, consumption and use of products*. Here again, the paragraph explicitly mentions the Circular Economy, which implies that these issues in particular can be regarded as building blocks for the Circular Economy transition.

6.3.1.3 The Circular Economy Package of 2014

The second Baroso Commission tried to fill the terminological and policy gap left by the Resource-Efficiency Initiative and the 7th EAP by the launch of a new and comprehensive policy programme in July 2014. This policy was meant to lay out a substantive strategy to establish a common and coherent EU framework for promoting the circular economy. The 2014 CE Package included a Communication and one legislative proposal that amends several waste directives all at once.[64] Taking into account the many references to other policies in the text of the 2014 CE Package, it was really meant to be the apotheosis of all the previous initiatives (the Resource-Efficient Europe and the 7th EAP being two major ones). It shows the relatively short period in which Circular Economy-related policies have merged into one common EU policy.

Not before long, however, the newly appointed European Commission officially withdrew the Circular Economy Package in December 2014 after just having been in office for less than two months.[65] While the 2014 CEP had been proposed in times of economic crisis, which could have restrained the Package's drafting in the first place based on prevailing economic and trade considerations, it was eventually withdrawn due to a wind of change that blew through EU politics. The Juncker Work Programme 2015, which announced the withdrawal, is officially called "A New Start" for a reason.[66] It stated, nonetheless, that the Commission wanted to propose a new, more ambitious proposal by the end of 2015 to promote a circular economy. Even though the Commission assured to come up with a more ambitious Package, it should be stressed that this was by no means legally binding, which left a lot of wiggle room for the Commission in terms of time, content and type of legal measure(s). Despite the ambiguity that surrounded the follow-up of the 2014 CE Package, neither the Council nor Parliament (or jointly) took the opportunity to challenge the withdrawal of the Package at the Court of Justice of the European Union.

---

64 European Commission, *Towards a circular economy: A zero waste programme for Europe*, COM(2014) 398 final/2; and European Commission, *Proposal for a Directive amending Directives 2008/98/EC on waste, 94/62/EC on packaging and packaging waste, 1999/31/EC on the landfill of waste, 2000/53/EC on end-of-life vehicles, 2006/66/EC on batteries and accumulators and waste batteries and accumulators, and 2012/19/EU on waste electrical and electronic equipment*, COM(2014) 397, respectively.

65 Ann. to European Commission, *Commission Work Programme 2015: A New Start*, COM(2014) 910.

66 European Commission, *supra* note 65.

At the end of the day, the European Commission kept its promise in the sense that it launched its "Circular Economy Package 2.0" in December 2015. The lead DGs for the Package are ENV and GROWTH. As pointed out in the introduction of this contribution, the CE Package now includes a Communication in the form of an Action Plan and several separate legislative proposals, each amending one particular EU waste directive,[67] such as proposals for the Waste Framework Directive and the Landfill Directive. The actions related to WtE are put forward in Section 6.3.4 of this contribution.

6.3.2 *The Definition of "Circular Economy"*

The Circular Economy Package contains a description of what is meant by the Circular Economy. The Commission explains in the introduction of the CE Action Plan that:

> the transition to a more circular economy, where *the value of products, materials and resources is maintained in the economy for as long as possible*, and the generation of waste minimised, is an essential contribution to the EU's efforts to develop a sustainable, low carbon, resource efficient and competitive economy. Such transition is the opportunity to transform [the] economy and generate new and sustainable competitive advantages for Europe.[68]

The Commission published a similar but more extensive definition of the Circular Economy transition in its withdrawn Circular Economy Package of 2014.

> Circular economy systems keep the added value in products for as long as possible and eliminate waste. *They keep resources within the economy when a product has reached the end of its life, so that they can be productively used again and again and hence create further value.* Transition to a more circular economy requires changes throughout value chains, from product design to new business and market models, from new ways of turning waste into a resource to new modes of consumer behaviour. This implies full systemic change, and innovation not only in technologies, but also in organisation, society, finance methods and policies. *Even in a highly circular economy there will remain some*

67 Note that these measures are the same ones that would have been changed by the legislative proposal of 2014.

68 Emphasis added. European Commission, COM(2015) 614, p. 2. Note that Commission Communications are policy documents with no mandatory authority. In simple terms, the Commission takes the initiative of issuing a Communication when it wants to give notice of its *own opinion* of a certain matter. A Communication has therefore no legal effect.

> *element of linearity as virgin resources are required and residual waste is disposed of.*[69]

Based on these descriptions,[70] the Circular Economy concept concerns the quest to create an economy in which *materials are used sustainably*. This includes that resources are kept within the economy when a product has become waste so that they can repeatedly be used productively and hence create further value. The materials system contains all materials in whatever form (as a resource, a substance, a product, waste…) and all activities related to materials. The approach not only aims at conserving natural resources, it also tries to reduce waste production and to minimize and control the environmental impacts of the materials we use throughout the running and following life cycles. A life cycle approach is really at the heart of the Circular Economy.[71] In light of this, the Circular Economy requires a high level of coherence between policies and legislation on materials and related policy areas, and, consequently, a high level of coordination between the stakeholders as well as their engagement and mobilization. It is furthermore worthy of note that the Commission acknowledges in the 2014 CE Package that even in a highly circular economy there will remain some elements of linearity. Consequently, practically speaking, there will probably always be *some* room for WtE activities. Overall, it is important to note that transforming the economy into a Circular Economy is the ultimate but also a hybrid goal; the choice of instruments that are used to steer the long-term transition depends on the particular situation, such as the mixture of materials that are used in the product, emerging technologies, the moment of action in the transition and which actor acts.

All things considered, the "Circular Economy" is comprehensive and complex, and therefore not a clear-cut concept; it is a collection of objectives and principles, which are generally shared with other similar concepts and initiatives. The CE Package proposed by the European Commission builds on these existing policies. In that perspective, the Flag-

---

69 Emphasis added. European Commission, *supra* note 64, p. 2.

70 Additionally, the Commission has posted a more extensive explanation of the Circular Economy on its website regarding the Questions & Answers of the CE Package (*see* <http://europa.eu/rapid/press-release_MEMO-15-6204_en.htm>, consulted on 4 November 2016). The Circular Economy is: "to ensure sustainable growth for the EU we have to use our resources in a smarter, more sustainable way. It is clear that the linear model of economic growth we relied on in the past is no longer suited for the needs of today's modern societies in a globalised world. We cannot build our future on a 'take-make-dispose' model. Many natural resources are finite, we must find an environmentally and economically sustainable way of using them. It is also in the economic interest of businesses to make the best possible use of their resources. In a circular economy the value of products and materials is maintained for as long as possible; waste and resource use are minimised, and resources are kept within the economy when a product has reached the end of its life, to be used again and again to create further value. This model can create secure jobs in Europe, promote innovations that give a competitive advantage and provide a level of protection for humans and the environment that Europe is proud of. It can also provide consumers with more durable and innovative products that provide monetary savings and an increased quality of life."

71 Life cycle thinking will be explained in more detail in the next section.

ship Initiative for a Resource-Efficient Europe and the 7th EAP can be seen as the formal stepping stones towards the Circular Economy Package.

6.3.3 *The Heart of the Circular Economy: Life Cycle Thinking*

The transformation from the current economy to a *Circular* Economy raises the question of what is meant by the word "circular". The term "circularity" is not explicitly explained in either of the Circular Economy Packages. What can be understood from the descriptions of the Circular Economy is that waste should not be seen as a problem, but rather as an opportunity: waste can be turned back into useable resources. Circularity, in that respect, means in any case the closing of one *material* life cycle and beginning a new one (or preferably even more than one cycle). The best known waste treatment for this conversional process is recycling.

A significant policy moment in this field stems from 2005 when the Commission published the Thematic Strategy on waste prevention and recycling.[72] In this Thematic Strategy the Commission expresses a real commitment to transform the EU into a "recycling society" by providing a number of measures that should stimulate recycling.[73] Although the Commission clearly links waste management to resources use, apparently it deliberately separates the Communication from another Communication on the general sustainable use of natural resources,[74] as both documents were released at the same date. This division can be explained by the fact that, at the time the Communications were drafted, waste had been at the centre of EU environment policy in the last 30 years.[75] The two Communications represent the previously ingrained perspectives on waste, i.e. that waste is a problem rather than an opportunity. It seems, however, that the Thematic Strategy on waste prevention and recycling nonetheless really tried to change this rhetoric by focusing on recycling and other recovery treatments, and thereby to perceive waste more and more as a valuable resource for industry. Key to that paradigm shift was the introduction of "life cycle thinking" to waste policy and legislation. Life cycle thinking takes into account *all* other stages of a material's life cycle in waste management and not only the "traditional" waste stage.[76] The Thematic Strategy on sustainable resource use

---

72 European Commission, *Taking sustainable use of resources forward: A Thematic Strategy on the prevention and recycling of waste*, COM(2005) 666.

73 *Id.* at pp. 17-20.

74 European Commission, *Thematic Strategy on the sustainable use of natural resources*, COM(2005) 670.

75 Note in this regard that the Flagship Initiative for a Resource-Efficient Europe was adopted only in 2011.

76 European Commission, COM(2005) 666, p. 7. The Review of the Thematic Strategy on the prevention and recycling of waste, also adopted in 2011, demonstrated, on the one hand, significant progress in the establishment and diffusion of the life cycle perspective in EU waste policy and law. On the other hand, it also signaled significant implementation differences between Member States of the binding minimum targets, which had been fixed in view of the life cycle approach under the 2005 Thematic Strategy, and made it quite clear that without further and complementary initiatives, opportunities will be missed to change the trends

carried out the same message as its fraternal twin policy on waste prevention and recycling: a systematic, coordinated and integrated application of the life cycle perspective was pursued, which cut-crosses policy fields.[77] What is important is that both Strategies focus on the materials system – *not* on the energy system.

As pointed out in Section 6.2.1.1, life cycle thinking found its way in the Waste Framework Directive of 2008 through the waste hierarchy. According to the non-legally binding Guidance on the Waste Framework Directive, the Commission describes life cycle thinking in the context of EU waste law as:

> a conceptual approach that considers upstream and downstream benefits and trade-offs associated with goods and services. [It] [...] takes into account the entire life cycle, starting with the extraction of natural resources and including material processing, manufacturing, marketing, distribution, use, and the treatment of waste. By introducing life-cycle thinking into waste policies, the WFD integrates waste policies, when departing from the waste hierarchy, into the broader framework of reducing environmental pressures and increasing resource efficiency.[78]

While the rational of life cycle thinking has slowly but surely been introduced to the policies and legislation on waste and resources, the concept originates in fact from the attempts of making *products* more sustainable.[79] The rise of reflecting a life cycle perspective in EU policy starts in 2001. The first Communication on the Integrated Product Policy (IPP) was the very first EU policy document to conceptualize the idea of life cycle thinking.[80] Back in 2001, the Commission started to rethink policies relating to the *overall environmental impacts* of products.[81] IPP can be described as:

---

and pressures expressed in Section 6.1.1. European Commission, *Report on the Thematic Strategy on the Prevention and Recycling of Waste*, COM(2011) 13, pp. 6, 8-9.

77 European Commission, COM(2005) 670, pp. 6-7.

78 European Commission, *supra* note 14, p. 49.

79 Decisions made in the period in which a product is conceptualized and manufactured are extremely important if products are to become more sustainable. It is estimated that more than 80% of all product-related environmental impacts plus the major costs throughout the product's whole life cycle are determined at the product's design stage. *See* BIO by Deloitte, Oeko-Institut and ERA Technology, *Preparatory Study to establish the Ecodesign Working Plan 2015-2017 implementing Directive 2009/125/EC – Task 1 Draft Final Report*, 2014, p. 5.

80 European Commission, *Green paper on integrated product policy*, COM(2001) 68.

81 While in some policy and legislation environmental impacts and human health impacts are separated, the environmental impacts include those on human health under the IPP approach. European Commission, COM(2001) 68, p. 2. Interestingly, the Communication did not include many economic or social considerations, even though the Commission stated that IPP plays an important role in contributing to Sustainable Development. According to the next IPP Communication, IPP is an integral part of the SDS. European

> an approach which seeks to reduce the life cycle environmental impacts of products from the mining of raw materials to production, distribution, use, and waste management. The driving idea is that integration of environmental impacts at each stage of the life cycle of the product is essential and should be reflected in decisions of stakeholders. IPP focuses on those decision points which strongly influence the life cycle environmental impacts of products and which offer potential for improvement, notably eco-design of products, informed consumer choice, the polluter pays principle in product prices. It also promotes instruments and tools which target the whole life cycle of products.[82]

This definition is important, because it lays the foundation for integrating environmental aspects into EU product policy and legislation, and shows that life cycle thinking is *not* based on considerations concerning the unsustainable production and use of energy. The 2001 IPP Communication lists plentiful concise guidelines for product design, including on the conservation of resources, the reduction of hazards and risks, and design for durability and recycling.[83] Overall, the Commission adopted a clear life cycle approach but without really defining it.

The subtitle of the sequel Communication on IPP (2003) was therefore no surprise: "Building on Environmental Life-Cycle Thinking".[84] The Commission considers life cycle thinking as a highly important "policy principle" in the 2003 IPP. It describes it as an approach that:

> considers a product's life-cycle and aims for a reduction of its cumulative environmental impacts – from the "cradle to the grave". In so doing it also aims to prevent individual parts of the life-cycle from being addressed in a way that just results in the environmental burden being shifted to another part. By looking at the whole of a product's life-cycle in an integrated way, IPP also promotes policy coherence. It encourages measures to reduce environmental impacts at the point in the life-cycle where they are likely to be most effective in reducing environmental impact and saving costs for business and society.[85]

Commission, *Integrated Product Policy – Building on Environmental Life-Cycle Thinking*, COM(2003) 302, pp. 2, 3 and 6.

82 European Commission, COM(2001) 68, p. 5.

83 *Id.* at pp. 19-20. In addition, suggestions were made on the generation of product information, labeling, Green Public Procurement (GPP) and standardization. *Id.* at pp. 12-13 and 15-21.

84 European Commission, *supra* note 81.

85 European Commission, COM(2003) 302, pp. 4 and 9.

Even without official policy follow-ups, the idea of taking the full life cycle of products into account to enhance environmentally friendly product policy and – importantly – corresponding policy coherence clearly entered the Union's industrial policy framework through the IPP Communications. The application of life cycle thinking has developed since then via the SCP Action Plan[86] to the Circular Economy Packages. Although the 2014 CE Communication features life cycle thinking to some degree, it is not very explicit: e.g. the layout is not in "life cycle order" and relatively too much attention is paid to the waste stage compared to other stages. In contrast, a number of life cycle stages are individually addressed in the 2015 CE Action Plan and they are even organized in consistence with the material life cycle to take the reader by the hand. In the 2015 version, there are strategic approaches proposed for four life cycle stages. In the nearby future, not only EU waste law will be modernized under the policy framework, but also, for example, the frameworks regulating products design and chemicals. In spite of receiving quite some criticism in the media for not being ambitious enough, these examples of differences between the 2014 and the 2015 CE Packages show the Commission's greater loyalty to applying life cycle thinking in the 2015 version than in the one published in 2014.

In conclusion, closing the life cycle and starting a new one through waste treatment practices such as recycling is not the only component of "circularity" that the Commission emphasizes in the CE Package of 2015 – albeit a very important one. Above all, the Commission stresses that the environmental impacts of the *entire* material life cycle should be considered in an integrated way.

Life cycle thinking – for which there is still no formal definition, making its application open for interpretation depending on the situation – is based on the idea that the use of materials causes a series of positive as well as negative environmental impacts during their life cycles. In view of that, it is important to search for opportunities to improve these effects in the life cycle stage(s) where it is best addressed, be it when materials are still raw materials, are used in products or are contained in waste. Even though WtE may be the best option in the waste stage, it is not actively stimulated because other recovery options "serve" the materials system better. This perspective is enclosed in the waste hierarchy.

86 European Commission, Communication on the Sustainable Consumption and Production and Sustainable Industrial Policy Action Plan, COM(2008) 397. The SCP Action plan builds on the IPP's interpretation of life-cycle thinking by aiming at a dynamic and integrated framework to improve the environmental (*this time expressly including energy!*) performances of *products* and to foster their uptake by users. In view of that, three priority areas are stressed which are also significant in the life-cycle perspective: resource-efficiency, environmental potential for industries and eco-innovation. The first policy field is addressed in the Resource-Efficient Initiative. The latter two policy fields are fleshed out in the Eco-innovation Action Plan (Eco-AP), which had been rolled out in 2011: European Commission, *Innovation for a sustainable Future – The Eco-innovation Action Plan (Eco-AP)*, COM(2011) 899.

6.3.4 *Proposed Changes to the Legal Framework Affecting WtE Operations*

The policy framework to address the transition towards a Circular Economy was launched by the European Commission in 2015 and is just as comprehensive as the term "Circular Economy". It includes an extensive Action Plan that lists both policy actions and legal actions and includes several legislative proposals for prominent directives in EU waste law. As should be expected and has already been highlighted above, the Action Plan follows a clear life cycle approach: for each life cycle stage a variety of actions is scheduled, most of which are short-term measures while aiming at long-term goals. Importantly, the life cycle stages are equally important: waste policy and legislation is therefore *not* the prime focus of the CE Package. The Commission additionally proposes extra actions targeting particular resource/product/waste groups. These priority areas serve as showcases of how to realize a Circular Economy. The Action Plan also throws light on the facilitating role horizontal measures play in the Circular Economy transition: besides the detailed actions concerning the life cycle stages and the priority areas, these conditions are key to trigger the systemic change aimed for.

Everything considered, law plays a substantial role in each of the three sections of the CE Action Plan. The legislative proposals, moreover, demonstrate how the Commission practically tries to reform the regulatory framework for the waste stage. The waste hierarchy remains very important in that respect. The actions and adjustments proposed in the legislative proposals that are significant for WtE practices, are set forth in the following sections.

6.3.4.1 Waste Hierarchy

The parts of the CE Action Plan concerning the waste stage and the "conversion stage" of the life cycle of a material both start by emphasizing that waste plays an important role in the Circular Economy, for waste management, particularly recycling, determines how the EU waste hierarchy is put into practice.[87] Clearly, both "circularity stages" are extremely important in the Circular Economy, as it converts waste into new materials useful to industry. The conversion from waste to resource should not only increase the security of resource supply in the EU, but should also lower the environmental and social impacts, which are typical for "primary" resource extraction. To facilitate the level-playing field in the internal market, action includes the simplification of cross-border formalities through research on material flows and the use of electronic data exchange. The Commission argues that a key factor in creating a dynamic market for secondary resources is the development of sufficient demand and that the role of the private sector in creating it is essential in that respect. Private initiatives should therefore be encouraged, the CE Action

87 European Commission, COM(2015) 614, pp. 8-13.

Plan states. Additionally, the use of public procurement policies is also mentioned as a method to create well-working markets for secondary materials. Actions to improve waste management are crucial, too, as proper waste practices have a direct impact on both the quality and quantity of the secondary resources used in the material cycle. The Commission expressly states that

> to achieve high levels of *material recovery*, it is essential to send long-term signals to public authorities, businesses and investors, and to establish the right enabling conditions at EU level, including consistent enforcement of existing obligations. All waste should be considered, be it generated by household, businesses, industry and mining [...], or the construction sector.[88]

Just as it is under the current Waste Framework Directive, the waste hierarchy is interpreted as an instrument that is first and foremost aiming at WtM.

6.3.4.2 Definitions

An important aspect of the CE Package is the terminology introduced to the Waste Framework Directive. New definitions are provided for "municipal waste", "construction and demolition waste", "final recycling process" and "backfilling". For example, backfilling, which falls under the "other recovery" category in the waste hierarchy, means:

> any recovery operation where suitable waste is used for reclamation purposes in excavated areas or for engineering purposes in landscaping or construction instead of other non-waste materials which would otherwise have been used for that purpose.[89]

6.3.4.3 Targets

According to the Commission, the rates for waste prevention, recycling and other forms of material recovery should be increased, as this would stimulate valuable materials finding their way back into the economy without (too many) harmful environmental impacts and significant economic losses. To accomplish this, the Action Plan highlights that it is essential to send long-term signals to public authorities, businesses and investors, and to establish the right enabling conditions at EU level. And that all waste should be considered, be it generated by household, businesses, industry, mining or the construction sector.[90]

---

88 Emphasis added. *Id.* at p. 8.

89 European Commission, *Proposal for amending Directive 2008/98/EC on waste*, COM(2015) 595, p. 14.

90 *Supra* note 89.

To accomplish the increase of waste prevention, recycling and other forms of material recovery, amongst several actions raised in the CE Action Plan,[91] the legislative proposals include several new and increased minimum targets for preparing for reuse and recycling. The target in the Waste Framework Directive for preparing for reuse and recycling of municipal waste is gradually increased from 60% by 2023 to 65% by 2030.[92] For C and D waste, the preparing for reuse and recycling target is increased to a minimum of 70% by weight and expanded to also include backfilling of non-hazardous construction and demolition waste.[93] The overall target in the Packaging Directive for preparing for reuse and recycling of packaging waste is increased to 65% by 2025. Moreover, EU-wide targets for paper and cardboard, glass, plastics, ferrous metal, aluminium and wood packaging are introduced: from 2025 to 2030; they are (almost) all gradually increased. For example, the target for wood packaging is increased from 60% in 2025 to 75% in 2030.[94] Last but not least, a binding landfill target is introduced in the Landfill Directive: the target reduces landfilling to a maximum of 10% of the total amount of municipal waste by 2030.[95]

6.4 Waste-to-Energy Communication: Where the Circular Economy Meets the Energy Union

The triangular relation between material, waste and energy is explicitly addressed in the CE Package. The CE Action Plan states that "action on the circular economy [...] ties in closely with key EU priorities, including [...] energy."[96] It further affirms that "'Waste to energy' can [...] play a role and create synergies with EU energy and climate policy".[97] To establish "how this role can be optimised, without compromising the achievement of higher [preparing for] reuse and recycling rates, and how the corresponding energy po-

91 Policy approaches such as raising awareness at national level, providing technical assistance to Member States facing difficulties, facilitating exchange of best practices between States and regions, and EU investments are examples mentioned in that regard. In addition, in order to stimulate high-quality recycling, both inside and outside the EU, the Commission commits itself to promote the voluntary certification of treatment facilities for certain key types of waste (*e.g.* electronic and plastic wastes). Unlike the other starting dates for the proposed actions (between 2015 and 2016 onwards), the promotion of industry-led voluntary certification of treatment facilities for key waste/recyclate streams only just begins in 2018. European Commission, COM(2015) 614/2, p. 3.

92 Lower targets are proposed for certain low-performing Member States, such as Greece, Croatia, Malta and Romania.

93 For changes made to the targets in the Waste Framework Directive see: European Commission, COM(2015) 595.

94 For changes made to the targets in the Waste Framework Directive see: European Commission, *Proposal for a Directive amending Directive 94/62/EC on packaging and packaging waste*, COM(2015) 596.

95 European Commission, *Proposal for a Directive amending Directive 1999/31/EC on the landfill of waste*, COM(2015) 594. A target is proposed for certain low-performing Member States, such as Greece, Croatia, Malta and Romania.

96 European Commission, COM(2015) 614, p. 2.

97 *Id.* at p.10.

tential can best be exploited,"[98] a word is also dedicated to the situation when waste cannot be recovered in view of the material cycle, that is the situation in energy recovery is preferable to disposal (which, as mentioned above, can also include WtE practices). To that end, the Commission planned to publish a Communication on how the Circular Economy can play a role in creating synergies between the framework regulating the materials system and the one for the energy system.

The Roadmap towards the Communication on Waste-to-Energy in the Circular Economy (WtE Communication) of 4 February 2016 states that:

> the Commission continues to work on a number of policies which converge towards the same objectives. [The WtE Communication] should seek synergies amongst them: the Circular Economy; the Energy Union Strategy; and the EU Climate Change and the Renewable Energy policies. Energy and material recovering non-recyclable waste not only makes sense under the resource circularity, but also would avoid landfill emissions with a powerful climate change potential; displace the use of virgin fossil fuels and contribute to the security of energy in the EU by producing energy from a renewable source.[99]

The Roadmap forms the starting point for the Waste-to-Energy Communication, as it already delineated the objectives and scope of the Communication.[100] The European Commission published the Communication "The role of waste-to-energy in the circular economy" at the end of January 2017.[101] On the whole, the Communication delivers what had been promised in the Roadmap, as it focuses on three issues:

1. clarifying the position of different WtE practices in the waste hierarchy and what this entails for public financial support;

98 *Ibid.*

99 European Commission, *Exploiting the potential of Waste to Energy under the Energy Union Framework Strategy and the Circular Economy*, ENV-UNIT A2-2016/ENV/086, pp. 1-2.

100 Most importantly, the Roadmap indicates five main challenges for WtE activities in the EU which would be addressed in the Waste-to-Energy Communication: 1. Lack of synergies between the WtE situation and EU policies; 2. Making existing WtE processes more energy efficient; 3. Unevenly spread WtE (over) capacities; 4. Untapped potential from waste-derived fuels; and 5. Lack of clarity with respect to the waste hierarchy. Id. at pp. 2-3.

101 European Commission, *The role of waste-to-energy in the circular economy*, COM(2017) 34. The official documents however indicated that the WtE Communication was to be published late 2016 (*see* European Commission, *Energy Union Package. A Framework Strategy for a Resilient Energy Union with a Forward-Looking Climate Change Policy*, COM(2015) 80; European Commission, *Annex to the Roadmap for the Energy Union*, COM(2015) 80/2, p. 4; European Commission, *An EU Strategy on Heating and Cooling*, COM(2016) 51; and European Commission, *supra* note 99. Mr Kestutis Sadauskas, who is Director at DG ENV of the European Commission announced that the Communication was rescheduled to early 2017. K. Sadauskas, International Solid Waste Association (ISWA), keynote speech at: "Circular Economy: the Role of Biowaste and Plastics", Brussels, 14 November 2016.

2. providing guidance to Member States on how to make better use of economic instruments and capacity planning; and
3. identifying the technology and processes that currently holds the greatest potential to optimize energy and material outputs.[102]

Regarding the first matter, the Commission underscores once again that the waste hierarchy, including the possibility to deviate from it, is the cornerstone of EU waste policy and legislation, and key to the Circular Economy transition. To support the transition, public financing of waste management, whether national or at EU level, should be consistent with the goal of moving upwards in the waste hierarchy. The Commission positions WtE practices on the hierarchy for the first time and recognizes that some WtE activities can be considered as "non-energy recovery", namely as other recovery forms or as disposal (or exceptionally as recycling).[103] Evidently, this has consequences for public financing: not only is the disposal of waste discouraged (which is by no means a novelty), Member States are also explicitly advised to gradually phase-out public support for the recovery of energy from mixed waste.[104]

With regard to the second challenge, the Commission emphasizes that the Circular Economy transition requires striking the right balance when it comes to WtE capacity for the treatment of non-recyclable waste, in that potential economic losses or the creation of infrastructural barriers to the achievement of higher recycling rates must be avoided. This basically ties in with the first issue highlighted in the WtE Communication. The Commission adds that for Member States with low incineration capacity and high reliance on landfilling cross-border waste shipments could help to make optimal use of the WtE capacity already available elsewhere in the EU, whereas for Member States with high incineration capacity recycling should be focused on a great deal more, amongst others through incineration taxes in addition to higher landfill taxes and a general freeze on the establishment of new WtE facilities and the shutting down of older and less efficient ones.[105]

And finally, the third issue stresses that due to the new rules on separate collection and the preparing for reuse and recycling rates as proposed in the Circular Economy Package, the amount of mixed waste available for WtE activities is expected to decline.[106] For other waste better collection, preparing for reuse and recycling rates are required. This, however, does not affect the general ambition to improve the energy-efficiency in

102 European Commission, *supra* note 101, p. 3. These three issues correspond to the five challenges of the Roadmap (*supra* note 100): the first issue corresponds to Roadmap challenges 1 and 5; the second issue corresponds to Roadmap challenge 3; and the third issue corresponds to Roadmap challenge 2.

103 *Supra* note 18.

104 European Commission, *supra* note 101, pp. 5-6. Note that mixed waste is a large part of the feedstock for WtE installations such as energy recovery incinerations.

105 *Ibid.*, pp. 6-8.

106 *Ibid.*, p. 8.

WtE processes, as this would maximize their contribution to the EU's climate and energy objectives.[107] The development of industrial symbiosis is an example, which is mentioned in this respect.[108]

The Commission seizes the opportunity in the concluding part of the WtE Communication to underline once again that WtE processes can most certainly play a role in the Circular Economy transition

> provided that the EU waste hierarchy is used as a guiding principle and that choices made do not prevent higher levels of prevention, reuse and recycling.[109]

6.5 Concluding Remarks

WtE activities may be considered to have two objectives: they manage waste and/or they generate energy. In point of fact, there are currently two historically grown, stand-alone regulatory frameworks for WtE at EU level, each predominantly aiming at one of those objectives: one framework regulating waste management (which is embedded in EU environmental law) and the other one regulating energy (which concerns EU energy law). Nevertheless, whether managing waste and generating energy are really distinct or contradictive objectives is debatable. One could argue, for example, that they go hand in hand and are in fact complementing each other. Recent developments in EU waste law may provide for change: the 2015 Circular Economy Package tries to establish the right regulatory framework for enabling the transition to a more Circular Economy in the European Union. This contribution studies the role WtE plays in this transition and throws light on what we can expect of the EU environmental framework for WtE activities in the future. Based on the previous sections of this contribution, one can conclude that *the Circular Economy transition is first and foremost a transition of the materials system – not of the energy system*. This conclusion can be drawn from the regulatory framework currently applicable to waste, the previous environmental policies underpinning both the existing waste framework as well as the CE Package, the CE Package as proposed by the European Commission in 2015 and, last but not least, the Roadmap to the WtE Communication.

There are many topics on the environmental agenda of the EU – waste is one of them. The waste hierarchy plays a key role in EU waste law, as it sets out a priority list for Member States to follow when drafting and executing national waste prevention and management measures. The hierarchy ranks three successive recovery categories (i.e.

107 *Ibid.*, pp. 9-10.

108 *Ibid.*, p. 10. Industrial symbiosis can be simply explained as an industrial area where a WtE plant processes the waste generated by the companies located in the area while providing them energy (e.g. heat or power) in return (*see also* Chapter 4, section 4.9.2 in this book).

109 *Ibid.*

preparing for reuse, recycling and a residual recovery category, within which energy recovery falls). Furthermore, it places all recovery operations before disposal operations. As pointed out earlier in this contribution, the waste hierarchy is based on life cycle thinking. If a Member State diverts from the hierarchy, this, too, is done on the basis of life cycle thinking if deviation delivers a better overall environmental outcome.[110] Depending on certain requirements and, in the case of recovering municipal waste for energy, certain energy-efficiency thresholds, WtE facilities can be seen either as recovery or as disposal (or even as recycling) facilities. Indeed, WtE activities can also sometimes be classified as disposal operations, such as is the case for waste incineration of municipal waste below the energy-efficiency thresholds and for the capturing and processing of gasses during the landfilling of waste. Keeping this in mind, one has to be careful when referring to "WtE" in situations where actually only energy recovery is meant,[111] which happens frequently, as this would pave the way for miscommunication and, more significantly, would have consequences for the proper application of the waste hierarchy. After all, some WtE practices, such as energy recovery and the reprocessing of waste into fuels, precede any disposal operation during which energy can still be generated (with low energy-efficiency, it must be said). As a general rule, the Waste Framework Directive favours WtM over WtE in the waste hierarchy, because it favours preparing for reuse and recycling over other recovery options and disposal activities respectively. Evidently, that does not mean WtE has no positive role to play as a *waste management treatment* in the Circular Economy: WtE activities prevent polluted or degraded materials from entering the recycling-process and re-entering the materials system again and divert waste from landfills.[112] The Commission does not deny this, as the definition of the Circular Economy provided in the 2014 CE Package recognizes that there will always be waste unfit for WtM processes. The WtE Communication, too, point this out.[113]

While the CE Package alters many provisions in several waste directives, it does not change the waste hierarchy. The Commission promised in the Roadmap to the WtE Communication no real changes to the policy and regulatory framework for WtE prac-

---

110 This also depends on the type of waste treated: for example, where some waste streams are best suited for recycling, others are best suited for energy recovery. K. Talus, 'Waste-to-Energy Through Gasification: An EU Law Assessment and Critique', *Oil, Gas, Energy Law Intelligence*, 14(3), 2016, p. 4.

111 A "WtE operation" (or plant/facility/installation) is a common way to refer to an energy recovery operation without making a clear difference between the WtE options based on the definitions and the waste hierarchy provided in the Waste Framework Directive, as discussed in Sections 6.2.1.1 and 6.2.1.2. They presume WtE operations to be energy recovery operations. *See e.g.:* M. Grosso, A. Motta and L. Rigamonti, 'Efficiency of Energy Recovery from Waste Incineration, in the Light of the New Waste Framework Directive', *Waste Management*, 30(7), 2010, pp. 1238-1243; and A. Evers, 'The Environmental Regulation of Energy-from-Waste Facilities in England', *Oil, Gas, Energy Law Intelligence*, 14(3), 2016, pp. 1-13 (although in this case clarification is provided later in the text on p. 7). Other authors point out which type(s) of WtE operation(s) they actually mean. *See e.g.:* Stengler, *supra* note 21, pp. 1 and 6.

112 Stengler, *supra* note 21, p. 4 and 13.

113 *See also* European Commission, *supra* note 101, pp. 4-5.

tices and has adhered to this promise: the waste hierarchy remained untouched and is even further embedded in EU environmental policy and law. The entire Communication reflects this idea.

The WtM-orientated approach is also reflected in the existing waste management targets: there are no targets for WtE operations in any form – there are, on the other hand, targets for the preparing for reuse and recycling. Note, moreover, that "material recovery" is expressly aimed for in the target for C and D waste in the Waste Framework Directive. By taking backfilling as a case in point, the Directive departs from the general rule that the operations falling under the residual recovery category of "other recovery" are on equal footing. These "clean" WtM targets do not only exist in the Waste Framework Directive, but appear across the entire legal landscape for waste, such as in the Packaging Directive and ELV Directive. Often, these targets are set parallel to targets for recovery in general (so including WtE recovery). Indeed, WtM recovery practices dominate the waste management targets (although disposal activities are discouraged through the new landfill target for municipal waste in the Landfill Directive, which would be beneficial for all recovery options). The CE Package does not drastically change this preference in the sense that it refocuses the waste policy towards energy generation. In fact, the Commission even reimburses its strategy by gradually increasing most of the targets in EU waste legislation, without adding new targets expressly aiming at WtE. Furthermore, the CE Package contains new definitions that illustrate the focus on the materials system, for example for "final recycling process" and "backfilling". The WtE Communication underscores the importance of the new waste management targets, amongst others by discouraging Member States to support energy recovery facilities.

The fact that the Commission does not deviate in the CE Package and in the WtE Communication from the WtM-orientated approach adopted in EU waste law can be explained by the gradual development of the Circular Economy transition in EU environmental policy and law. The concept of life cycle thinking has been around for some time now, laying down the foundation for the Circular Economy transition. Based on previous environmental policies, it can be concluded that life cycle thinking, which has in the first place been developed in product policy whereupon it has been implemented in waste and resource policies, is meant to address the circularity of *materials*. And so, even before the Circular Economy Package was officially launched, circularity had been progressively advocated, both in policy and in legislation. It makes therefore good sense not to change the waste hierarchy but, instead, to put even more emphasis on waste prevention, preparing for reuse and recycling. For a while now, the implementation of life cycle thinking in the regulatory framework for the entire materials system has already been set in motion, and, as we can see, is now accelerating under the umbrella of the Circular Economy Package. Against this background, one could argue that we can speak of "EU materials law" whenever the entire regulatory framework for the Circular Economy is meant. This "new" legal field covers the full life cycle(s) of materials. It covers the regulations on resources for the technical sphere, chemicals, products and waste accordingly. Consequently, EU waste law is not the only field targeted by the CE Package; directives and regulations regulating other life cycle stages are also aimed at. The issues addressed in that respect are, for example, the design of products and the public procurement of products and construction. These issues, too, are addressed in the CE Package and chiefly concern aspects relevant to the materials system, such as durability and reparability. The role WtE plays in the development of these "other" life cycle stages of the Circular Economy Package is trivial in comparison to the aim of delivering full circularity in the materials system.

The strong belief that environmental considerations should be embedded in other policy fields than only the environmental field is apparent in the case of the Circular Economy transition. Despite the fact that environmental concerns lay at the origin of the Circular Economy movement, the 2015 CE Action Plan puts strong emphasis on sustainable economic growth.[114] For example, it highlights the creation of new jobs, opportunities for new business, innovations and investments, and the stimulation of resource-efficiency and competitiveness.[115] Amongst other reasons, the comprehensiveness of the CE Package can be explained by the strategy chosen by DG ENV. DG ENV allied with different Directorates to gather support for the 2014 as well as the 2015 CE Package, e.g. DG for Employment, Social Affairs and Inclusion[116] and DG GROWTH, which includes the divisions Internal Market and Industry. In point of fact, DG GROWTH is even the other lead DG, together with DG ENV. In addition to the Circular Economy goal of policy integration, which derives both from the transitional and the live-cycle perspective adopted in the CE Package, working together with other DGs was politically speaking arguably the best option, because DG ENV alone would possibly not have been able to get enough support for the Package.[117] It is noteworthy, moreover, that DG ENV did *not* work together (at least not publicly) with the DG responsible for the Energy Union, Energy and Climate Action.[118]

---

114 Both the Resource-Efficiency Europe Initiative and the 7th Environment Action Programme confirm this. *See e.g.* Recital (7) and Article 2(1)(b)-(c) Decision 1386/2013 of 20 November 2013 on a General Union Environment Action Programme to 2020 "Living well, within the limits of our planet", OJ L 354/171.

115 Besides the clearly visible economic-driven objectives, the Action Plan also points at the *wider* environmental benefits of a Circular Economy, i.e. that a Circular Economy would save energy and help avoid the irreversible damages caused by resource use in terms of climate and biodiversity, air, soil and water pollution. In other words: benefits that fall outside the direct scope of the materials system. European Commission, COM(2015) 614, p. 2.

116 When the 2014 CEP was launched by DG ENV, DG Employment, Social Affairs and Inclusion simultaneously presented a new initiative, called the Green Employment Initiative. European Commission, Green Employment Initiative: Tapping into the job creation potential of the green economy, COM(2014) 446.

117 Arguably, the close liaison between both DGs may have contributed in the choice not to call it the "Circular Society", even though that term would have been adequate considering the society-wide scope of the concept.

118 The new DG ENER merges several former DGs, such as DG CLIMA and DG ENER, into one general DG, and adds the newly launched ambition of the EU to create a so-called "Energy Union" as well.

In spite of choosing this strategy, the strong belief that environmental considerations should be embedded in other policy fields, *including* the energy framework, and thus the need to optimize synergies and to address the trade-offs between these different policy fields, is enshrined in the integration principle (Article 11 TFEU). Put differently, life cycle thinking and the overall ambition of establishing a Circular Economy in the European Union ought to be integrated in (politically speaking) non-environmental policies and regulations, such as the ones concerning the energy system (hence, even though DG ENER was not directly involved in the drafting process of the CE Package). In fact, this already happens in some cases: the waste hierarchy and life cycle thinking in general are not only key to EU materials law, they actually also play a role in energy legislation. For example, the Renewable Energy Directive (RED)[119] includes binding targets for the utilization of renewable energy by 2020. It establishes that Member States can encourage the use of biofuels made from waste (Preamble 89 RED) and includes elements of industrial and municipal waste into the definition of biomass (Article 2(e)), and hence establishes that biodegradable waste is a renewable energy resource. The Directive refers to concepts contained in the Waste Framework Directive in several instances, such as to the definition of waste (Article 2(p)) and to the waste hierarchy (Article 3(5)). Based on the applicability of the integration principle and on these existing connections, it may seem at odds that the responsibility of the Circular Economy Packages is not – to say at least – partly shared with DG ENER. The publication of the WtE Communication does not change the situation.

The link between the Circular Economy and other policy and regulatory frameworks may be found at different points in the Circular Economy Package. The Circular Economy concept touches by its very nature on several policy fields, such as food supply, land use, climate control and energy supply and use. While the Circular Economy Package is more straightforward in addressing the materials system, the Resource-Efficiency Initiative and the 7th EAP are not unambiguously clear on the exact boundaries between the materials system and the energy system. "Resources" are interpreted quite broadly in both policies. On the other hand, resource-efficiency includes delivering greater value of products with less input in the materials system. This shows that the Circular Economy is in any case highly interconnected to other sustainability transitions and that these transitions should be pursued in parallel.[120] Together, these transitions can be considered as the pillars to establish an all-inclusive sustainable economy in the EU. This also means that the aim of establishing a Circular Economy in the European Union is not a completely stand-alone exercise – waste is not exclusively the domain of the materials system.

119 Directive 2009/28 of 23 April 2009 on the Promotion of the Use of Energy from Renewable Sources and Amending and Subsequently Repealing Directives 2001/77/EC and 2003/30/EC, OJ L 140/16.

120 Another interrelated transition is, for example, the one for agriculture and food. In that transition, bio-waste plays a role as well, as it provides for opportunities for bio-waste based fertilizers. For the materials system, on the other hand, bio-waste can be used in the production process of bio-materials, such as bio-plastics used in packaging.

The existing waste framework regulating WtE practices supports this consideration. That said, it can be argued that the Circular Economy transition and the energy transition should still be seen as two separate transitions. Ultimately, if the EU economy really transforms into a so-called Circular Economy by 2050, where ever more waste is prevented from being produced and more and more waste is prepared for reuse and is recycled, there will be less and less "residual" waste left for WtE practices, in the form either of energy recovery or of disposal. For this reason, waste cannot play a *decisive* role in the transition to a more sustainable energy system in the long run (*albeit* beneficial for the energy transition in the short run). What would be interesting to see is what the WtE Communication will actually bring about.

In conclusion: based on existing EU waste law, the policy roots of the Circular Economy Package, the Package itself and the WtE Communication, the Circular Economy transition is above all a quest to transform the materials system, amongst others through law. It means that EU materials law, which addresses not only the waste stage but also other life cycle stages, will be adapted and strengthened in view of this goal. WtE is regarded as a positive side-effect of the Circular Economy, and in that way merely helping the energy transition.

7 Regulating Emerging Technologies: Regulation and Innovation in WtE

Kim Talus and Topi Turunen

7.1 Introduction

There are various ways to convert waste to energy: combustion of waste in mass-burn incinerators, gasification, pyrolysis and plasma processes, anaerobic digestion for gas production, biomass facilities using low-grade feedstocks, methane extraction from landfill, etc. Depending on the specific circumstances, these can be seen as alternatives and the selection of the method depends on the details of each case. Available waste feedstock, planned power production capacity, possible pre-existing power production facility and other parameters will impact the choice. Various alternatives are still being developed and against this background, it is important that the regulatory framework regarding waste-to-power technology is technology-neutral.

Similarly, different methods of producing power and heat should be placed on a level playing field. What should be the focal point is the emissions. This would contribute to the simplicity of the regulation and, at the same time, would ensure that the environmental objectives are not being compromised. Regulation should not pick the winners, instead of markets.

This chapter includes a technical terminology for different waste streams applicable for WtE production: first of these is "municipal solid waste" or "MSW", which consists of household waste items discarded by the public.[1] The composition of MSW varies greatly depending on the geographical area and changes significantly over time.[2] Therefore, it is

---

1 *See* COM (2015) 595 final. Proposal for a Directive of the European Parliament and of the Council amending 2008/98/EC on waste, p. 12: In the draft proposal to make amendments to the WFD, municipal waste is defined as (a) mixed waste and separately collected waste from households including: paper and cardboard, glass metals, plastics, bio-waste, wood, textiles, waste electrical and electronic equipment, waste batteries and accumulators, bulky waste, including white goods, mattresses, furniture, garden waste, including leaves and grass clipping; (b) mixed waste and separately collected waste from other sources that is comparable to household waste in nature, composition and quantity and (c) market cleansing waste and waste from street cleaning services, including street sweepings, the content of litter containers, waste from park and garden maintenance. Municipal waste does not include waste from a sewage network and treatment, including sewage sludge and construction and demolition waste. So even with the amendment, MSW consists of really heterogenic waste streams.

2 *See*, for example, Öztürk, Tuba; Veli, Sevil; Dimogno, Anatoly, 'Issue of municipal solid waste and the process of its management in solution', in Machado, Claudio & Konno, Gianni (eds.), *Municipal Solid Waste* (Nova Science Publishers Inc. 2011), pp. 41-52, at p. 41.

impossible to survey its characteristics and consistency on a wide scale. Due to its heterogenic consistency, MSW fits poorly in material recovery operations, and therefore it is a cheap waste stream with a large production volume to use in energy production, which is essential for WtE production.

MSW is seldom burned without further processing and it is often processed into "refuse-derived fuel" or "RDF" or "solid recovered fuel" / "specified recovered fuel" or "SRF" materials.[3] RDF and SRF are produced by shredding and dehydrating MSW so that energy can be produced more efficiently. The shredding of the MSW is essential so that WtE plants can utilize the material without risk of blockages being caused by bigger pieces of MSW. Dehydration is carried out to make the incineration process more efficient due to the decreased humidity of the solid waste material. SRF is essentially the same material as RDF but with a quality standard, often making it high-quality RDF.[4] Standards for SRF are set in standard EN15359.[5] The consistency and features of RDF are less predictable and RDF is mostly used as a general term to describe MSW that has been properly prepared for efficient WtE operation.[6]

"Producer gas" is the last important concept discussed in this chapter. The main difference between producer gas and natural gas is that, producer gas is manufactured from another product and natural gas is gathered from natural sources. Producer gas can be manufactured from materials such as wood, coal and coke or other carbonaceous material, but can also be manufactured from waste-based substances. References in this chapter to producer gas should be understood as producer gas manufactured from waste-based substances such as MSW, RDF or SRF. The emissions produced by burning producer gas are different from those produced by burning natural gas: for example, burning producer gas may produce heavy metal emissions due to the chemical composition of the waste substances involved contrary to burning natural gas. Producer gas can be produced in many different ways, among which thermal treatment is one of the most common.

This chapter will focus on the EU law treatment of one technology: producer gas. This gas is often used in pre-existing utility boilers as additional fuel. In its current state, EU law effectively prevents or at least significantly complicates this type of activity, regardless of its potential benefits to EU energy and waste policy objectives.

EU regulation should support any measures through which emissions can be reduced. The overall logic of EU emission control is that it favours an approach where emission

3 *See* European Commission: Reference Document on the Best Available Technique for Waste Incineration. Integrated pollution Prevention and Control 2006, p. 7; Monte, M.C.; Fuente, E.; Blanco, A. & Negro, C., 'Waste management from pulp and paper production in the European Union', *Waste Management* 29 (2009), pp. 293-308, at pp. 300-301. Pure MSW can also be used in energy production but it is not as efficient because of its moisture, etc.

4 BREF 2006, pp. 222-223: The identifiability provided by the standards assists in the recovery of materials.

5 Standard provided by the European Committee for Standardization.

6 The definitions of RDF and SRF vary between different countries. In this chapter these terms refer to the materials defined in this paragraph.

limits are set for power plants. This sustainability objective can be achieved by setting specific emission standards for plants using producer gas. Where the primary fuel is being partially replaced by producer gas, the emissions can be reduced. Compared to more traditional mass-burn incineration, the use of producer gas has much higher efficiency and through it even over 60% more electricity can be generated.[7] Against this, the regulatory choice to place further limits on producer gas, compared to combustion of SRF, must be questioned. While SRF can be excluded from waste regulation through a national (or potentially EU level) end-of-waste regulation, producer gas is being subjected to a double regulatory threshold: the "normal" end-of-waste regulation as well as a comparison to natural gas emissions, means that producer gas is treated differently from other waste-to-power applications.

This chapter will focus on waste incineration, a process of turning waste into useful energy: heat and/or power. By focusing on the Industrial Emission Directive (2010/75/EU)[8] and its regulation of gasification of waste, it shows how bad legislative drafting can cause significant problems for both the government licensing authorities and the economic operators in this area.

7.2 Waste Incineration and EU Energy and Waste Policy Objectives

The objectives of EU energy and waste policies are not that different. EU energy policy aims at reducing both the use of natural resources and curving European dependence on imported energy as well as reducing greenhouse gas emissions.[9] EU waste policy strives for example to reduce the amount of waste generated and to maximize recycling and reuse.[10] From the point of view of these objectives, EU waste and energy policies can be seen as complementary. Energy recovery can be economically efficient and higher and higher efficiency rates can be achieved, especially through gasification.[11] The utilization of waste-based replacement fuels decreases the environmentally harmful impacts of energy production when compared to coal, for example, and reduces the usage of raw

7 A. Read, 'Why energy from waste?', in A. Evers and A. Read (eds.), *Energy from Waste: A Practical Handbook* (Globe Law and Business 2015). *See also* Commission Press Release, State aid: Commission authorises investment aid for "energy from waste" gasification plant in Lahti, Finland (IP 14/108, Brussels 4 February 2014).

8 Directive 2010/75/EU of the European Parliament and of the Council of 24 November 2010 on industrial emissions (integrated pollution prevention and control) Text with EEA relevance (OJ L 334, 17.12.2010, p. 17-119).

9 Generally on EU energy policy objectives, *see* Talus, Kim, *EU Energy Law and Policy: a Critical Account* (Oxford University Press 2013).

10 Decision No 1386/2013/EU of the European Parliament and of the Council of 20 November 2013 on a General Union Environment Action Programme to 2020 'Living well, within the limits of our planet' (OJ L 354, 28.12.2013, pp. 171-200).

11 This was noted in a Commission Press Release (IP 14/108, Brussels 4 February 2014).

natural resources and the need to import energy, it can be argued that waste-based replacement fuel production is supported by both EU waste and energy policies.

7.2.1 Waste Incineration in the Context of EU Energy Policy

EU energy policy objectives evolve around the "energy trilemma": security of energy supply, sustainability and competitiveness. The complementary nature of EU waste and energy policies was apparent in the recent State Aid decision from the European Commission for a waste gasification plant in Lahti, Finland (Kymijärvi II project), which suggested the obvious – gasification of waste supports both EU energy and environmental goals. In that case, European Commission assessed the measure under the EU environmental aid guidelines, and found that

> gasification allows for a cleaner and more efficient recovery of energy from waste than traditional techniques. In particular, it allows for a more efficient generation of energy and a better power-to-heat ratio. The most important impact of the investment is that the previously mostly fossil energy based heating will be much more environmental friendly, especially in terms of CO2-emissions which are expected to decrease by 279.000 tons per year. The plant also contributes to increasing security of energy supply and diversifying energy supply sources. In addition, the project will reduce the need to use landfill sites for future waste, generate knowledge for research and improve sustainable waste management.[12]

As can be seen, according to the views of the European Commission and its DG Competition, gasification promotes EU energy policy objectives: energy efficient generation (an objective of the Directive 2012/27/EU on energy efficiency), security of supply and reduction of import dependency (a key energy policy objective for natural gas), reduction of CO_2 emissions (another key objective for EU climate and energy policy), sustainable waste management and reduction of landfill waste (key objectives for waste policy in the EU). Clearly, gasification of waste presents an opportunity to address the policy objectives of both EU energy and climate policy and waste policy.

Given these linkages between the two policy area objectives, it is no surprise that it has also been noted as one area of interest for the current Energy Union activities:

12 *Idem.*

> The Commission will further establish synergies between energy efficiency policies, resource efficiency policies and the circular economy. This will include exploiting the potential of "waste to energy".[13]

A communication in this respect was released in January of 2017. The communication discusses the problems of WtE sector in energy production and as an instrument of waste management especially focusing on the balancing of the energy production capacities of WtE operations in different Member States as well as the rightful position of WtE in the waste hierarchy. In addition, the efficient ways of recovering energy such as gasification of quality waste streams (namely SRF) from waste are emphasized.[14]

7.2.2 *Waste Incineration in the Context of EU Waste Policy*

7.2.2.1 Waste-based Fuels and the EU Waste Hierarchy

Directive 2008/98/EC on waste[15] (the "Waste Framework Directive") provides for a general framework of waste management requirements and the basic waste management related definitions for the EU. The main aim of the Waste Framework Directive (WFD) is to create a functioning "circular economy" within the EU where the amount of end-of-life waste would be kept to a minimum and waste would be used as raw material in industry as an alternative to using new raw materials thus moving from "a cradle to grave" approach to "a cradle to cradle" approach.[16] A circular economy of this kind would have significant economic benefits as well as reduce the need for, and environmental impacts of, landfills.

The Circular Economy Package[17] includes a number of references to the idea that waste can serve as an energy source within the economy. According to the Circular Economy Action Plan, the transition to a Circular Economy is an essential contribution to the EU's efforts to develop a sustainable, low-carbon, resource-efficient and competitive economy. The Circular Economy pursues the ideal situation where the value of products, materials and resources is maintained in the economy for as long as possible and

13 Communication from the Commission, A Framework Strategy for a Resilient Energy Union with a Forward-Looking Climate Change Policy (COM/2015/080 final).

14 Communication from the Commission, The role of waste-to-energy in the circular economy (COM/2017/34 final).

15 Directive 2008/98/EC of the European Parliament and of the Council of 19 November 2008 on waste and repealing certain Directives (OJ L 312, 22.11.2008, pp. 3-30).

16 Communication from the Commission to the Council, the European Parliament, the European Economic and Social Committee and The Committee of the Regions – Taking sustainable use of resources forward – A Thematic Strategy on the prevention and recycling of waste (COM(2005) 666 final), pp. 3-4, 6.

17 The Package includes an Action Plan: European Commission, *Closing the loop – An EU action plan for the Circular Economy*, COM(2015) 614.

where the generation of waste is minimized.[18] A clear life cycle perspective is adopted in the Action Plan; various policy and regulatory actions are planned for each life cycle stage for the coming years. The Commission Communication on the challenges regarding WtE practices examines these ideas further.[19]

The definition of waste has had a great impact on waste management in general, and forms a cornerstone of European waste legislation.[20] The application of the waste legislation in its entirety depends on the definition, since such legislation applies only to operations that produce or treat waste. While this has for a long time only been subject to the case law of the Court of Justice of the European Union,[21] the Waste Framework Directive now regulates the possibility to remove substances from the legal classification of waste (or define the by-products of the substance in question, which are not legally classifiable as waste) at EU and at the national level.[22] For example, solid recovered fuel (SRF) has been removed from the classification of waste in certain EU Member States and are therefore regarded in legal terms as fuel products.[23] The end-of-waste process will be discussed in more detail in the following sections.[24]

The Waste Framework Directive lays down measures to protect the environment and human health by preventing or reducing the adverse impacts of the generation and management of waste and by reducing the overall impact of resource use and improving the efficiency of such use. Article 4 of the Waste Framework Directive introduces a waste hierarchy, which applies in relation to waste prevention and management legislation and policy at national and EU level and comprises of:

a. prevention;
b. preparing for reuse;
c. recycling;
d. other recovery, e.g. energy recovery; and
e. disposal.[25]

18 European Commission, COM(2015) 614, *ibid.*, p. 2.

19 De Römph, Thomas – Reins, Leonie: Waste-to-Energy and the Circular Economy – Connecting the Dots. OGEL Waste-to-Energy special issue 2016 (<www.ogel.org>).

20 Communication from the Commission: Towards a thematic strategy on the prevention and recycling of waste (COM(2003) 301 final), p. 22. The definition of waste has been examined in much detail in G. Van Calster, EU waste law (OUP 2015).

21 *See*, for example, Joined Cases C-418/97 and C-419/97, *ARCO Chemie Nederland and Others* [2000] ECR I-4475 and Case C-9/00, *Palin Granite and Vehmassalon kansanterveystyön kuntayhtymän hallitus* [2002] ECR I-3533.

22 For the application of Article 6 of the Waste Framework Directive, see Case C-358/11, *Lapin Luonnonsuojelupiiri* (ECLI:EU:C:2013:142).

23 For example, Italy and UK. Both examples are further discussed below.

24 This has also been discussed in T. Turunen and K. Talus, 'National regulatory possibilities in respect of solid recovered fuel', *Renewable Energy Law and Policy Review* (2/2014).

25 Traditionally, waste regulation has emphasized disposal.

The waste hierarchy is not legally binding. However, it is clear that where possible, the waste hierarchy should be respected. The current waste hierarchy imposes more concrete obligations than its prior version.[26] However, many of its obligations should be considered as imposing a duty to use "best efforts" than actually to achieve particular results.[27] In most cases, it is enough for Member States to strive only to achieve the objectives of the hierarchy.[28] Thus leaving the effectiveness of the hierarchy depending on national and EU-wide implementation measures. The objective of the hierarchy is to reduce the amount of waste produced and encourage the reuse and recycling of waste instead of exploiting new natural resources.[29] As such, the underlying idea is the promotion of the above-mentioned waste hierarchy and the idea if circular economy. At first site, it would appear that waste incineration does not align itself with the waste hierarchy. However, a closer look will show that this is not necessarily so. First, the different nature of different waste streams must be considered in the application of the waste hierarchy. Whereas some waste streams are best suited for recycling (such as certain metals or glass), others are best suited for energy recovery (certain plastics, processed wood, foodstuff styrofoam and plastics, etc.).[30] Second, circular economy should also be understood to mean "more energy from less waste".[31] So a circular economy approach should not necessarily be understood as an approach aiming at "no waste" but rather as an attempt towards minimum waste and as a "waste to resources" approach where waste is turned into products and energy. By using the most advanced technology, it is possible to extract maximum amounts of energy from the waste. The above-mentioned State Aid decision in the *Kymijärvi II* case indicates a similar approach: "the project will reduce the need to use landfill sites for future waste." In practice, these efficiencies will come from new technologies such as pyrolysis and gasification.

The use of waste as a feedstock for energy production is increasing internationally. This is driven by various factors including the growing volumes of waste being generated, combined with the fact that even at best, landfills are a suboptimal solution for potentially valuable material. This is also the starting point in the EU where the EU Waste Frame-

26 Directive 2006/12/EC. Directive 2006/12/EC of the European Parliament and of the Council on waste (OJ L 114, 27.4.2006, pp. 9-21).

27 Best efforts obligations permit the Member States to resort to excuses when they cannot or do not comply with the objectives.

28 Van Kempen, Jasper, 'Obligations of result or best efforts: Dealing with problems of interpretation', in Peeters, Marjan & Uylenburg, Rosa (eds.), *EU Environmental Legislation: Legal Perspectives on Regulatory Strategies* (Edward Elgar 2014), pp. 146-172 at pp. 149-156.

29 Turunen, Topi and Van Calster, Geert, '*Burning Construction and Demolition Waste: An Assessment of the European Waste Hierarchy and Recovery Targets,*' OGEL 14(3) (2016) Waste-to-Energy (WtE) special issue, pp. 1-5.

30 The current plan is that low-grade wood and plastic would come out of the conventional waste hierarchy and be used for energy recovery. For details, *see*: <http://european-biogas.eu/2015/06/03/circular-economy-to-exploit-advanced-efw-technology/>.

31 <http://european-biogas.eu/2015/06/03/circular-economy-to-exploit-advanced-efw-technology/>.

work Directive places landfills at the bottom of the waste hierarchy.[32] Efficient use of waste for energy production is a clear advantage over the landfill option. Another driver is the energy potential of certain waste streams.[33] From an energy policy perspective, energy security, rising energy prices,[34] decentralized energy production from local fuel and the need to cut emissions – all point towards waste-based energy generation.[35]

In European waste legislation, all operations are considered as either recovery operations or disposal operations. Placing WtE operations into one of these two categories has a decisive effect on the operation of installations and the legislative framework governing them.[36] Both operation categories are defined in Article 3 of the WFD. Article 3(15) defines recovery as "any operation the principal result of which is waste serving a useful purpose by replacing other materials which would otherwise have been used to fulfil a particular function, or waste being prepared to fulfil that function, in the plant or in the wider economy." Article 3(19) defines disposal operations as "any operation which is not recovery even where the operation has as a secondary consequence the reclamation of substances or energy." The main difference between the categories appears to be that in recovery operations the waste serves a useful purpose and fulfils a beneficial function as in disposal operations it serves no such purpose.

The efficiency of the WtE installation seems to be the crucial issue in separation between the categories: since waste always produces energy in an incineration process, it is only a matter of efficiency as to whether it produces enough energy to be considered to fulfil a beneficial function in a plant or in the wider economy. In recovery operations waste produces enough energy to be considered a substitute for virginal energy production fuel, while in disposal operations waste may produce some energy but the main objective of the operation is to dispose of it.

In numerous cases, the Court of Justice of the European Union (CJEU) has dealt with the classification of WtE operations. One of the principles it developed at an early stage was that in recovery operations waste has to replace raw materials and thus help conserve

32 Despite this, landfills continue to be the primary waste disposal option in many Member States. For details, *see* <www.mater.polimi.it/mater/en/news-events/news/456-waste-to-energy-the-new-coolsweep-report> (last visited 12.6.2015).

33 This is also recognized by the EU Commission, which has indicated that part of the future circular economy plans is to generate more energy from less waste. "This supported exploiting the most advanced technologies, such as pyrolysis and gasification, 'to the full'." In practice, this means that low-grade wood and plastic could come out of the conventional waste hierarchy and be used for energy recovery. For details, *see*: <http://european-biogas.eu/2015/06/03/circular-economy-to-exploit-advanced-efw-technology/>.

34 The currently low energy prices are largely a result of low demand, unconventional oil and gas in the United States and similar transitory factors and will most likely be a short-term phenomena. For example, *the US shale oil fields are expected to start their decline as soon as the early 2020s. See* <www.iea.org/publications/freepublications/publication/WEO_2014_ES_English_WEB.pdf>.

35 Read, 2015.

36 *E.g.* Van Calster, Geert, 'Waste management in the EU', in Van Calster, Geert; Vandenberghe, Wim & Reins, Leonie (eds.), *Research Handbook on Climate Change Mitigation Law* (Edward Elgar 2015), pp. 361-372, at p. 365.

natural resources.[37] In *Commission v. Germany*[38] the CJEU held that incineration was to be classified as a recovery operation when it fulfilled the criteria set out in paragraph R1 of Annex II of the directive on waste then in force.[39] These were as follows: (1) the amount of energy produced must be larger than the amount of energy consumed in the operation; (2) the surplus energy produced should be used in producing heat or electricity; and (3) the waste should be mainly used as a fuel or other means of producing energy.[40] In *Commission v. Netherlands*[41] the amount of produced energy and its feasibility was considered in relation to a WtE operation. A great deal of subsequent case-law has also referred to the R1 efficiency standards or R1 formula contained in Annex II of the WFD.[42]

A WtE operation is either a R1 operation where the waste is used principally as a fuel or other means to generate energy and is considered a recovery operation, or a D10 operation where the waste is incinerated on land and is a disposal operation. The D10 operation is listed together with other examples of disposal operations in Annex I of the WFD. The main distinction between R1 and D10 is that if the D10 operation produces energy at all, the production should be so inefficient that it cannot be considered the main function of the operation. The rationale behind D10 is waste disposal. WtE installations that can produce energy in an efficient manner should be considered R1: recovery operations. The present precise R1 efficiency formula for burning MSW can be found in the footnote to paragraph 1 of Annex II of the WFD.

7.2.2.2 Waste-based Replacement Fuels and End-of-Waste Regulation

Article 6 of the Waste Framework Directive provides for rules on the so-called "end-of-waste status" and regulates the circumstances under which certain specified types of waste can cease to be treated as waste. End-of-waste status is reached when the waste has undergone a recovery operation, including recycling, and complies with specific criteria[43] to be developed at EU or national level in accordance with the following cumulative conditions:

37 C-6/00, *Abfall Service* (2002) ECR I-1961.

38 C-228/00, *Commission v. Germany* (2003) ECR I-1439.

39 Directive 75/442/EEC. Council Directive 75/442/EEC on waste (OJ L 194, 25.7.1975, pp. 39-41).

40 *See* Evers, Angus, 'Environmental regulation of energy-from-waste facilities', in Evers, Angus & Read, Adam (eds.), *Energy from Waste: A Practical Handbook* (Globe Law and Business 2015), p. 135. *See also* "Guidelines on the interpretation of the R1 energy efficiency formula for incineration facilities dedicated to the processing of municipal solid waste according to Annex II of the WFD", European Commission 2011.

41 C-113/02, *Commission v. Netherlands* (2004) ECR I-9707.

42 C-116/01, *SITA v. VROM* (2003) ECR I-2969, C-458/00, *Commission v. Luxembourg* (2003) ECR 1153, C-251/07, *Gävle Kraftvärme AB* (2008) ECR I-7047.

43 In addition to the requirements of Article 6 of the Waste Framework Directive, it is necessary that the requirements developed in the case law of the CJEU are respected. For example, it is required that the owner of the waste does not discard it or intend or is required to discard it. *See* Joined Cases C-418/97 and C-419/97, *ARCO Chemie Nederland and Others* [2000] ECR I-4475, Case C-9/00, *Palin Granite and Vehmassalon*

a. the substance or object is commonly used for specific purposes;
b. a market or demand exists for such a substance or object;
c. the substance or object fulfils the technical requirements for the specific purposes and meets the existing legislation and standards applicable to products; and
d. the use of the substance or object will not lead to overall adverse environmental or human health impacts.

A substance or object can cease to be waste through fulfilling the cumulative end-of-waste criteria laid out above. In certain Member States some standardized WtE fuels have already ceased to be waste on the strength of these criteria.[44] At the level of the EU has been opposition towards WtE materials ceasing to be waste but the national possibilities to regulate still remain, as long as they respect the criteria set in Article 6.[45]

Since the MSW-based material has a very specific purpose, it seems evident that the first criterion is fulfilled in WtE operations. In addition to that, a large number of energy plants can utilize MSW, RDF and SRF in energy production in a profitable way: there is a clear market and demand for the use of waste in energy production. The crucial issue in respect of the first two end-of-waste criteria is to ensure that end-of-waste is not simply a means of circumventing the waste disposal legislation.[46] The focal question with the third criterion is that it is possible to use the material from a legal point of view after it ceases to be waste and waste legislation stops applying to it. After the waste legislation stops applying, all the other legislation that would apply to a similar product starts to apply.[47] Hence, it must be ensured that the end-of-waste product can actually be used in the same way as its pristine counterpart.[48] The fourth criterion extends the objectives of the third criterion to ensure that using material as a non-waste does not cause any adverse environmental or human health impacts. The criterion imposes that the object or substance cannot cause more environmental or human health effects after it ceases to be waste compared to the situation when it was still waste.[49] The impact assessment must take into account the life

kansanterveystyön kuntayhtymän hallitus [2002] ECR I-3533 and Case C-358/11, *Lapin Luonnonsuojelupiiri* (ECLI:EU:C:2013:142).

44 *E.g.* Italy, UK.

45 *See* CMTD (2010) 0880: The Committee for the adaptation to scientific and technical progress and implementation of Directive 2008/98/EC on waste.

46 JRC Scientific and Technical Reports: End of Waste criteria, final, European Communities 2008, p. 8. However, if the need for heat is low in the country where the operation is located, the market situation might be somewhat more difficult since WtE operations are rarely profitable if they only produce electricity.

47 *E.g.* In the case C-358/11, *Lapin luonnonsuojelupiiri*, the CJEU ruled that the characteristics of the CCA saturated poles were regulated thoroughly enough via Regulation (EC) No. 1907/2006 (REACH).

48 *See* JRC Scientific and Technical Reports: Study on the selection of waste streams for end-of-waste assessment, Final Report. European Commission 2010, p. 36.

49 JRC Scientific and Technical Reports: End of Waste criteria, final, European Communities 2008, pp. 10, 16, 32.

cycle of the product and the positive effects resulting from the material ceasing to be waste.

It appears that when it is possible to control the environmental and human health effects of the WtE operations in a sufficient way, MSW-based fuels could attain end-of-waste status. However, due to MSW's heterogeneity, not just any MSW going to energy production can cease to be waste: the consistency of the fuel and the emissions caused by incinerating MSW can be erratic.[50] Regardless, if the consistency and the emissions can be controlled, there are no barriers to end-of-waste status. The standards for SRF might work as an instrument in ensuring the quality of MSW-based energy production fuels and the impacts of the WtE installation. The crucial factor is that in the absence of quality and emission control in incineration, MSW-based fuels cannot cease to be waste. The standard would not have to be the EN-standard for SRF: it could be any standards under which adverse environmental and human health impacts can be avoided.

Article 6(4) of the Waste Framework Directive provides that where no end-of-waste criteria have been set at EU level, "Member States may decide case by case whether certain waste has ceased to be waste taking into account the applicable case law."[51] No EU-wide end-of-waste criteria exist for waste-based replacement fuels, such as SRF. Therefore, the enactment of national provisions laying down end-of-waste criteria for SRF are an available option for Member States if the criteria laid down in Article 6 are fulfilled.

Regardless of whether the end-of-waste criteria are set at EU or national level, Member States must comply with the cumulative conditions laid down in Article 6. These primarily aim to ensure that an object or material previously defined as waste will be utilized in its new function and not merely disposed of.

The Italian regulation on SRF is based on the premise that SRF meets the cumulative conditions set out in Article 6.[52] The market demand for SRF is relatively clear, but the Italian legislation also provides for other necessary conditions relating to the environment and human health. The relevant provision states that it is:

---

50 Arena, Umberto, 'Process and technological aspects of municipal solid waste gasification. A review', *Waste Management* 32 (2012) pp. 625-639, at p. 637. Manufacturing producer gas from pure MSW seems to involve the same problem of heterogeneity as affects WtE operations in which solid waste is burned.

51 For case law on the definition of waste and its interpretation, *see*: C-235/02, *Saetti* (2004) ECR I-1009; C-418/97 & 419/97, *ARCO Chemie* (2000) ECR I-4475; C-188/07, *Commune de Mesquer v. Total France SA & Total International Ltd.* (2008) ECR I-4501; C-444/00, *Mayer Parry* (2003) ECR I-6204; C-114/01, *AvestaPolarit Chrome* (2003) ECR I-8752; C-9/00, *Palin Granit* (2002) ECR I-3533; C-206-207/88, *Vessosso & Zanetti* (1990) ECR, I-01461; C-1/03, *Van de Valle et al.* (2004) ECR, I-07613; C-129/96, *Inter-Environnement Wallonje* ASBL(1997) ECR, I-7435; C-422/92, *Commission v. Federation of Germany* (1995) ECR I-1124.

52 The English version is available at <http://ec.europa.eu/enterprise/tris/pisa/app/search/index.cfm?fuseaction=getdraft&inum=1836604>. The document is also available in Appendix I to this document.

> necessary to promote the production and use of Solid Recovered Fuels (SRF) to be used, under certain conditions, as a replacement for traditional fuel for environmental and economic purposes with the aim of contributing to reducing polluting emissions, including climate-altering gasses, to increasing the use of renewable energy sources through sustainable use of the biomass contained in waste for energy purposes, with a higher level of waste recovery, in compliance with the waste hierarchy pursuant to Article 179 of Legislative Decree no. 152 of 3 April 2006, to reducing the environmental and economic burden of waste disposal in landfills, saving natural resources, reducing dependence on traditional fuels, and increasing the certainty of the power supply.[53]

There are also other Member States with varying end-of-waste regulation or guidance for waste-based fuels. These would include UK where the national guidance explicitly makes a reference to Article 6 of the Waste Framework Directive.[54]

7.3 Industrial Emissions Directive and Producer Gas

The Industrial Emissions Directive provides for the rules on integrated prevention and control of pollution arising from industrial activities. It also lays down rules designed to prevent or, where that is not practicable, to reduce emissions into air, water and land and to prevent the generation of waste, in order to achieve a high level of protection of the environment taken as a whole.

Following the judgment from the CJEU in C-317/07, *Lahti Energia Oy*,[55] a provision, Article 42 (1), was added in the Industrial Emissions Directive. Article 42(1) provides that

> 1. This Chapter shall apply to waste incineration plants and waste co-incineration plants which incinerate or co-incinerate solid or liquid waste.
>
> This Chapter shall not apply to gasification or pyrolysis plants, if the gases resulting from this thermal treatment of waste are purified to such an extent that they are no longer a waste prior to their incineration and they can cause emissions no higher than those resulting from the burning of natural gas.

53 Available at <http://ec.europa.eu/enterprise/tris/pisa/app/search/index.cfm?fuseaction=getdraft&inum=1836604>.

54 For the UK example, *see* <https://www.gov.uk/government/uploads/system/uploads/attachment_data/file/341718/LIT_9992.pdf>.

55 C-317/07, *Lahti Energia Oy* (2008) ECR I-9051.

Where the producer gas (produced through gasification or pyrolysis) meets the criteria set in Article 42(1), Chapter IV, which provides for the special provisions for waste incineration plants and waste co-incineration plants, does not apply to the operation of the gasification and pyrolysis plant, nor to the main power generation plant. The application of this provision is critical for situations, for example, where an existing coal-fired power plant considers investing into a gasifier that would enable gasification of waste and the use of producer gas to replace a degree of coal in the process. In this situation, the new investment risks turning the coal-fired power plant into a waste incineration or waste co-incineration plant. This has severe consequences for the operation of the power plant and makes the operation of the plant technically and commercially much less appealing, especially in periods of lower heat and power demand. Other problematic issues in the application of Article 42(1) will be discussed below.

While the interpretation of Article 42(1) could at the outset be considered quite simple and straightforward, a more careful examination shows that the provision is both difficult to interpret, and does therefore not provide legal certainty for the economic operators in this area, and is fundamentally flawed in its rationale. It is of paramount importance that these shortcomings are rectified. Regulatory support for waste-to-energy conversion is crucial for the industry. For example, the absence of strong regulatory drivers diverting waste from landfills in the United States or Australia has meant that the waste-to-power applications have not been adopted as widely as in the EU (although the picture is slowly changing in both countries).[56] The absence of a predictable regulatory framework for gasification in the EU, will undermine the further development of a "more efficient option for waste-to-energy plants in the EU".[57]

7.3.1 *Origins of Article 42 (1) of the Industrial Emissions Directive*

It appears that Article 42(1) of the Industrial Emissions Directive has a connection with the decision of the CJEU in case C-317/07, *Lahti Energia Oy*. In that case, the Court had the opportunity to assess the interphase between waste and producer gas. The Court noted quite specifically that the qualities of the producer gas could be such that it can be assimilated with fossil fuels:

> 35. In that connection, first, as it appears from the information given by the referring court, the gas concerned, by reason in particular of its filtration in the purifier, has properties similar to a fossil fuel and thereby constitutes a gas suitable for use as a fuel for the production of energy both in the power plant for which the production of the gas plant is intended and in other power plants.

56 Read, 2015.

57 Commission Press Release (IP 14/108, Brussels 4 February 2014).

> 36. In those circumstances, there is no question of a substance resulting from the thermal treatment of waste in the gas plant which is incinerated in the power plant in order to complete a simple process of waste disposal. As the Finnish and Italian Governments have submitted, when the process is completed within the gas plant a product having the characteristics of a fuel is generated from waste.

The Opinion of Advocate General Kokott[58] is even more explicit:

> 56. As with otherwise ascertaining whether a substance is waste, the question whether a substance ceases to be waste must be determined in the light of all the circumstances. It is relevant, for example, whether there is a market for the processed product and whether the power plant could obtain on the market a fuel possessing comparable characteristics which is not regarded as waste. However, marketability alone is not sufficient, because even substances and objects with a commercial value can be waste. Instead, the important factor is whether the processed waste possesses characteristics comparable to the corresponding primary raw material or the corresponding product, in particular as regards environmental risks.
>
> 57. There is good reason to assume that, prior to filtering, the producer gas is not yet sufficiently similar, on account of impurities, to other products or primary raw materials, whilst the purified producer gas is possibly comparable to natural gas and similar gaseous fuels. Ultimately it is for the referring court to examine this point.
>
> 58. The answer to the fourth question must therefore be that the purified gas generated at the gas plant may be regarded as a product, so that the rules on waste no longer apply to it, if it is sufficiently similar to primary raw materials or other products.[59]

As such, both the Court and the AG suggest that it is possible that producer gas has the same qualities as a fossil fuel. This suggests that a comparison must be made. The Court does not provide guidance for deciding on the reference fuel for the purposes of carrying out the comparison. This, of course, is in line with the approach of the Waste Framework Directive and the end-of-waste process. The comparable fuel is to a certain extent dependable on the alternative fuels or the replacement fuel. In her Opinion, AG Kokott

58 Opinion of AG Kokott in C-317/07, *Lahti Energia Oy*, delivered on 11 September 2008.
59 *Ibid.*

suggests that "purified producer gas is possibly comparable to natural gas and similar gaseous fuels."

The focal point in this comparison between purified producer gas and other comparable fossil fuels must, quite naturally, be on emissions ("environmental risks," as AG Kokott puts it). AG Kokott makes the reference to natural gas and similar gaseous fuels. It is important to note in this context that there are various gaseous fuels that can be used for industrial purposes; these include industrial gas from steel or petrochemicals industries or landfill gases. Natural gas is but one fossil fuel that has a gaseous form. Instead of selecting natural gas as the only or even primary comparable, the emphasis in the decision and the Opinion is that the properties of the new product must be sufficiently similar to comparable fossil fuels.

The approach of the Court is in line with the rationale of the end-of-waste regulation in the EU. However, for reasons that are not entirely clear, it seems that the legislator reacted to the Court's decision in *Lahti Energia Oy* by adding a provision (Article 42(1)) to the Industrial Emissions Directive that goes significantly further than the Court's decision. Unlike the underlying decision and Opinion, the new article does not refer to comparable products or such, but instead makes a reference to natural gas emissions only.

This choice has certain unfortunate consequences. Importantly, it narrows the comparable to natural gas emissions only, and fails to pay heed to the fact that producer gas is not natural gas. This question will be examined next.

7.3.2 *Interpretation of Article 42(1) of the Industrial Emissions Directive*

According to Article 42(1) of the Industrial Emission Directive,

> 1. This Chapter shall apply to waste incineration plants and waste co-incineration plants which incinerate or co-incinerate solid or liquid waste.
>
> This Chapter shall not apply to gasification or pyrolysis plants, if the gases resulting from this thermal treatment of waste are purified to such an extent that they are no longer a waste prior to their incineration and they can cause emissions no higher than those resulting from the burning of natural gas.

Applying Article 42(1) in practice involves significant uncertainties. A literal interpretation, which should be the starting point for any interpretation of the EU's *acquis*,[60] re-

60 *See, for example*, cases C-263/06, *Carboni* [2008] ECR I-1077, para. 48 and C-48/07, *Les Verges du Vieux Tauves* [2008] ECR I-10627, para. 44. A contextual interpretation also supports the same two-step-logic. On the interpretation of EU law in general, *see* 283/81, *Cilfit* [1982] ECR 3415.

veals that the provision introduces two producer-gas-specific requirements. In order to be excluded from the application of Chapter IV, the producer gas must be purified to the extent that: (1) it ceases to be waste prior to the incineration; and (2) it cannot cause emissions that are higher than those resulting from burning of natural gas.

In line with this two-step logic, the first requirement seems to suggest the application of the EU end-of-waste regulation to producer gas. The new product, producer gas, must therefore have similar qualities and usage as the replacement product.

Waste-based producer gas can include certain natural impurities, which can be transmitted to air, land or water. These impurities must be eliminated through purification before such gas can acquire end-of-waste status. This process is regulated through Article 6(1) of the Waste Framework Directive, under which four cumulative conditions must be met. Article 6(1) also provides that the criteria to be developed in accordance with these conditions should include limit values for pollutants where necessary and take into account any possible adverse environmental effects of the substance. These limit values must be set by using a suitable replacement fuel, which could be a commonly used gas.

Without going into detail as to possible forms of end-of-waste regulation for SRF[61] or producer gas, it may be noted that the main focus at this stage should be on those impurities that are not eliminated as a part of the normal process of purifying flue gas. In this way, the new product, producer gas, can be used in the same way as the replacement fuel and without any new purification systems being introduced within the power-producing facility. Similarly, the focal point in the end-of-waste stage of the application of Article 42(1) should be product qualities, not emissions. Emissions are then examined in the context of the natural gas comparison at stage 2 of the assessment.[62]

While no EU level end-of-waste regulation for SRF or product gas has yet been enacted, end-of-waste regulation for some of these fuels has been put in place in certain Member States, such as the UK[63] for example. Here, the comparison seems to be with replacement fuel.

The second requirement – that product gas cannot cause emissions which are higher than those resulting from burning of natural gas – requires comparison of the emissions. Under Article 3 of the Industrial Emissions Directive, "'emission' means the direct or indirect release of substances, vibrations, heat or noise from individual or diffuse sources in the installation into air, water or land." This definition, quite logically, refers to an emission released from the industrial plant, such as a power plant. Therefore, this means that the point of measurement is "at the end of the chimney". The consequence of this is

61 Italy and the UK provide examples on this point.

62 It is also necessary to note that the emissions are also regulated in the permitting stage of the power plant.

63 See <https://www.gov.uk/government/uploads/system/uploads/attachment_data/file/341718/LIT_9992.pdf>.

that the emissions limit must be plant-specific. This, again, is rather logical.[64] Not only is this interpretation in line with the wording of Article 42(1), but it also corresponds to the general objective of the Industrial Emission Directive[65] and the more general objective of EU environmental regulation: prevention and reduction of emissions.

The comparison with natural gas emissions is not without complications. In essence, producer gas is not natural gas. The composition of these two types of gases is different, and they are accordingly not entirely comparable. Therefore, any comparison must be made *mutatis mutandis* and the natural differences between the two must be considered as part of the comparison.

To be clear, it is not commercially possible to turn producer gas into natural gas. Certain differences in the composition of these two gases will always remain. As a result, this requirement of the Directive cannot be followed literally. A further complication is the absence of limit values for certain components in natural gas. The EU regulation for natural gas does not provide for any limit values for certain substances (heavy metals, chlorine or dioxin, for example). When Article 42(1) provides for a comparison to be made without clarifying how the comparison should be made in these cases, the problem is evident. An interpretation relying on an approach where the lack of a limit value for a component in the IE-Directive should be considered to mean zero (0) emission of that component, is untenable. More commonly, unspecified limit values mean unlimited emissions, which are not subject to any controls. For example, the burning of coal will cause minor emissions of heavy metals and Cl. However, there are no limit values for these components in the IE-Directive. This is premised on the fact that the emissions of these components are minor, not that they are not present at all.

Given that various fuels have different compositions, and emissions of specific particles will differ accordingly, the IE-Directive sets different emission limit values for different fuel types. These differences must be considered and the limit values must be set so as to ensuring high level of environmental protection. The key issue is to provide a numerical value that creates certainty for the industry and prevents environmental dumping.

The applicable EU law should not seek to assimilate two different products, just as it should not treat two competing products differently – unless there is an objective reason to treat different products alike, or similar products differently.[66] Consequently, the Industrial Emissions Directive cannot require a full comparison between the two different gases. Instead, the natural differences in their composition must be recognized and taken into consideration. This approach would also be in line with that taken in the *Lahti Energia Oy* case.

64 A simple comparison to illustrate this is gasoline and cars. The emission limits in that context are not set for the gasoline itself, but for the exhaust fumes.

65 *See* Article 1 of the Industrial Emission Directive.

66 On discrimination, *see*, for example, C-17/03, *VEMW and others* [2005] ECR I-4983.

The next section will now recap and outline the interpretative difficulties that the current wording of the IE-Directive poses for both the licensing authorities working with the waste-to-power projects and the investor looking to invest in this technology.

7.3.3 Interpretative Difficulties with Article 42(1) of the IE-Directive

7.3.3.1 First Difficulty: Departing from the General EU Approach

The first difficulty with the current wording of Article 42(1) is that by making the direct comparison with natural gas, the article departs from the more general practice and logic in setting emission standards for various fuels in the European Union (EU). It also causes direct discrimination of producer gas and the producers of this type of gas.

The general approach in the EU, including that of the IE-Directive, is to set fuel-specific emission limits. While the merits of this type of fuel-specific approach can be questioned from the environmental perspective, it is nevertheless the EU approach. While failing to provide for fuel-specific emission standards for producer gas, the EU departs from its standard approach.

Producer gas is not natural gas. The composition of these two types of gases is different. As such, they cannot be fully compared. In equating the two very different gases, Article 42(1) discriminates against producer gas. Any comparisons between the two must be made *mutatis mutandis* and the natural differences between the two must be considered as part of the comparison. Clearly, emission standards for natural gas are designed for natural gas, not for other types of gases. The approach for other industrial gases like blast furnace gas, coke oven gas or low calorific gases from gasification of refinery residues has been to create gas-specific requirements.

7.3.3.2 Second Difficulty: Unclear Wording of Article 42(1)

The second problem with the current wording of Article 42(1) relates to its interpretation. The unfortunate wording of the provision makes is unclear whether the comparison to natural gas should be read as requiring direct comparisons to product qualities of natural gas or the emissions caused by burning of natural gas. This is a significant practical problem for licensing authorities and has a detrimental effect on investments in this area. Given the numerous benefits from producer gas for both energy and climate policy objectives of the EU,[67] this is unfortunate.

If the provision is interpreted as requiring a comparison for product qualities of the two gases, the reading of the article runs into two significant problems. The question of discrimination has been discussed above: producer gas is not natural gas. As such, these two different fuels cannot be directly compared. More importantly, the rationale of the

67 For discussion, *see* Commission Press Release (IP 14/108, Brussels 4 February 2014).

provision is to ensure a high level of environmental protection. Specifically, it aims at ensuring that emissions from waste incineration are kept under control and below certain standards.[68] This also applies to the producer gas versus natural gas comparison: the rationale is to limit emissions. Emissions from any fuel are fully dependent on the technology used in converting the energy of the fuel source into heat and/or power. The environmental impact of burning of natural gas in an open flame, as is widely done in oil fields where associated natural gas can be flared, is detrimental to the environment. As such, the focal point should be on emissions, not the product qualities of these fuels.

While the only logical conclusion is that the reference to natural gas must be to emissions, not product characteristics, this difficulty in interpretation has caused unnecessary interpretative difficulties.

A further difficulty with the current wording of Article 42(1) is that the reference to natural gas is linked to possible emissions from natural gas, not to any specific natural gas emission limits, not even to those provided in the annexes to the IE-Directive. Given that emissions from burning of natural gas are highly user- and technology-specific, this lack of precision is unfortunate. This creates uncertainty for both licensing authorities and economic operators applying for permits.

7.3.3.3 Third Difficulty: Lack of Emission Standards for Natural Gas

Even if the wording of Article 42(1) would be changed to ensure clarity of interpretation and, as a result, licensing authorities and economic operators would have certainty that the reference is to the emissions of natural gas as provided in the annexes of the IE-Directive, the next difficulty arises from the fact that emission standards set for natural gas focus on certain emissions (SO_2, NO_x, CO and dust particles).

Other components, though some of these can be present in natural gas in small quantities, but are more significant for producer gas, have not been considered as requiring limit values for natural gas use (such as heavy metals and chlorine). Because of the product quality differences in producer gas and natural gas, there is a need to create limit values for those components.

An interpretation relying on an approach where the lack of a limit value for a component in the IE-Directive should be considered to mean zero (0) emission of that component, is untenable. More commonly, unspecified limit values mean unlimited emissions, which are not subject to any controls. For example, the burning of coal will cause minor emissions of heavy metals and Cl. However, there are no limit values for these

68 According to Article 1 of the IE-Directive: "This Directive lays down rules on integrated prevention and control of pollution arising from industrial activities." Also, according to preamble 34 of the IE-Directive: "In order to ensure a high level of environmental and human health protection [...], it is necessary to set and maintain stringent operating conditions, technical requirements and emission limit values for plants incinerating or co-incinerating waste within the Union."

components in the IE-Directive. This is premised on the fact that the emissions of these components are minor, not that they are not present at all.

Given that various fuels have different compositions, and emissions of specific particles will differ accordingly, the IE-Directive sets different emission limit values for different fuel types. These differences must be considered and the limit values must be set so as to ensure a high level of environmental protection.

7.3.3.4 Fourth Difficulty: Fuel-specific Emission Limits and Licensing

Given the obscure wording of Article 42(1) and the absence of any specific emission limit values for producer gas, the IE-Directive fails to provide rules on how to alter the environmental permit and provide for emission standards for the new mix of fuels, in the event that a waste gasifier is added into an existing power plant that has been licensed for fossil fuels, such as coal, or for biomass. The absence of specific regulation creates uncertainty for both licensing authorities and the economic operators in this area and has a detrimental impact on investments.

7.3.3.5 Fifth Difficulty: End-of-Waste Criteria for Producer Gas

As said, the current wording of Article 42(1) provides that

> This Chapter shall not apply to gasification or pyrolysis plants, if the gases resulting from this thermal treatment of waste are purified to such an extent that they are no longer a waste prior to their incineration and they can cause emissions no higher than those resulting from the burning of natural gas.

While the provision makes a clear reference to the end-of-waste process under Directive 98/2008/EC, it does not specify criteria for this process. One possible interpretation has been to understand the provision as making a reference to natural gas as part of the end-of-waste process. Due to the reasons explained above, this interpretation cannot be sustained.

Clearly, there are significant interpretative difficulties with Article 42(1) of the IE-Directive. As such, there is a need to rethink the wording of the provision and bring the wording in line with the objective of this particular provision and the more general objective of the IE Directive.

7.4 Finally

Markets should pick winners, not governments. What the government must do, however, is to control negative externalities: pollution. As such, the government regulation in the area of waste-based energy production should be on emissions. This approach would treat all technologies equally and in line with the more general regulation of power generation facilities. The power plant specific limit values provided for in the Industrial Emissions Directive, specified in the national legislation implementing the Directive, as well as in the licensing of power plants, must be respected. Controls over quality of producer gas must be tight and rigorously followed. This ensures not only the best environmental outcome but also the use of only those technologies that can deliver a stable and high-quality product.

The logic of EU waste and energy policy objectives both support the high-efficiency waste-to-energy conversion provided through gasification of certain waste streams. As such, the regulatory support for this activity at EU level should be a "no-brainer" for the policy makers. Or at least regulatory hurdles should be eliminated.